Draw logs from Dowsville...

The History of the Ward Lumber Company

Draw Logs from Dowsville...
The History of the Ward Lumber Company

Mary A. Gow
Kitty Werner

DISTINCTION PRESS
WAITSFIELD, VERMONT

Draw logs from Dowsville…The History of the Ward Lumber Company
Mary A. Gow & Kitty Werner

Published by Distinction Press
354 Hastings Rd
Waitsfield, VT 05673
www.distinctionpress.com

Copyright © 2011 Mary A. Gow, Kitty Werner and Distinction Press

Present Day Photography © 2011 Kitty Werner

Designed by Kitty Werner, Distinction Press

All rights reserved.

ISBN 978-1-937667-04-7

Cover Photo: the original Upper Mill on the Mad River, circa 1930s

Our many thanks to the people of the historical societies who contributed their time and photographs to this work:

Duxbury Historical Society
Moretown Historical Society
Skip Flanders of the Waterbury Historical Society
University of Vermont Special Collections
Vermont Historical Society, Barre
Vermont State Library, Montpelier

Rights and Permissions

Interviews

Vermont Folklife Center, Middlebury

All sidebar quotations, unless otherwise noted, are from the 1992 Mad River Valley oral history project of the Vermont Folklife Center, based in Middlebury, VT. http://www.vermontfolklifecenter.org/

The selections are from the following taped interviews.

p.56	Lettie Conrad 1992.0049
	Robert Wimble 1992.0047
	Bob Gove 1992.0057
	Mary Reagan 1992.0059
	Frena Cutler 1992.0048
	Wilma Maynard 1992.0060
p. 58	Robert Wimble 1992.0047
	Frena Cutler 1992.0048
p. 88	Frena Cutler 1992.0048
p. 89	Bob Gove 1992.0039
	Gerard Dunbar 1992.0038
p. 90	Nelson Patch 1992.0032
	Clesson Eurich 1992.0005
	Robert Wimble 1992.0047
p. 101	Guy Livingston 1992.0016
	Warren White 1992.0046

Photographs

Page 8—Eastman Business School: Photo courtesy The Early Office Museum, www.officemuseum.com

Pages 19 and 20—Lowell, MA mills on Merrimack River

Credit Line: Library of Congress, Prints & Photographs Division, Detroit Publishing Company Collection, [reproduction number, e.g., LC-D4-10865]

Page 80—Vermont Department of Forests, Parks and Recreation, Forestry Centennial http://www.vtfpr.org/htm/for_cen_history.cfm

From the UVM Special Collections: noted as such with each photo

From the Moretown Historical Society: noted as such

From the Duxbury Historical Society: noted as such

Photographs taken in 2011 by Kitty Werner

All other photos and unattributed newspaper clippings are from the Ward family collections

Dedication

This book was created at the request of Owen M. Ward, and developed with both Owen and Holly Ward, not only to remember the history of the Ward Lumber Company, but to honor all the people associated with it who made the company as successful as it was over its century of history. They should not be forgotten.

"If it weren't for our employees, we wouldn't have been able to do what we did."

— Owen Ward, 2011

Acknowledgement

We thank Owen Ward for this opportunity to learn about and record the history of the Ward Lumber Company, and sincerely appreciate all his help with this project. It was been a great pleasure to work with someone so thoughtful, intelligent, and knowledgeable about his family's business and the industry.

We also wish to thank Holly Ward for his endless help with family photos, treks to old mill sites, demonstrations at the current clapboard mill, the tour at Lamell Lumber and many phone calls patiently explaining details of mill operations. Working with Holly has been quite the treat, as well as an education in Vermont history.

To those dedicated people who devote their time and energy to saving our collective history, our great appreciation and thank yous: Skip Flanders of the Waterbury Historical Society, Nadia Smith at the UVM Special Collections, Don Welch and Eulie Costello of the Duxbury Historical Society, Denise Gabaree of the Moretown Historical Society, the staff at the Vermont Historical Society and the Vermont State Library.

Our thanks to Mary Murphy for finding us and her continued support, to Tom O'Keeffe and Lamell Lumber Company of Essex, Vermont for the grand tour that helped us understand the entire process from cut logs to finished lumber, the Vermont Folklife Center for their awesome tapes reminding us, too, of old friends now departed, the Vermont Department of Forests, Parks and Recreation for the help and knowledge—we couldn't have done without you.

—*Mary A. Gow and Kitty Werner*

Foreword

On behalf of the Ward Family I would like to thank everyone who helped make this book a reality. Many have participated, and many have contributed in different ways over the past 15 years since my father, Owen Ward, and I first discussed the idea of a simple compilation of his "memoirs." That early seminal thought began to take form when, during a family reunion many years ago, I asked Dad to dictate his memories as we drove through the Mad River Valley. It had been many years since the family had all been together in Vermont following the untimely death of our mother, Jean Francis Brate Ward. During the first part of the trip Dad was reluctant to share his history, but the familiar surroundings seemed to inspire him as the words began to flow. You could see in his eyes the memories flow like the river itself as we drove down Route 100, at times calm and peaceful, and then suddenly turbulent or sad as he recalled the rough and rocky sections. It was clear that life in the early days of Vermont was unpredictable. It was obviously filled with joyful family memories, but as in life, it was also filled with setbacks and sometimes heartache. I am inspired by the resiliency and optimism of the Ward Family and all the other early Vermont pioneers as they faced enormous challenges, setbacks, successes and failures.

After I got home from that trip my secretary transcribed a bag full of cassettes, and it became perfectly clear that this was a much bigger task than we first envisioned. The many pages of rough and garbled dictated notes gathered dust in my office for many years while life marched on. I knew the task would emerge again. Two winters ago I was having lunch with Dad (When I am in Florida I have lunch every Wednesday with Dad. Much like the book *Tuesday's with Morrie*, I was having "Wednesday's with Owen"!), and on one of those Wednesdays Dad brought up the subject of the book again. Suddenly and emphatically it was clear that he had decided that it was time to finish the job! His persistence from that point forward was so characteristic of him and his ancestors and the dogged determination with which they pursued their goals. Without him, this book would never have been written!

We knew we needed professional help to write this book, so we gathered up old photos, documents, newspaper articles, or anything we could find that might be of interest, and we put it all in the capable hands of Mary Murphy and asked her to assist us in finding the right people to help. Mary organized the materials and identified a number of local authors and professionals that might be of assistance.

We were very fortunate to end up in the hands of Mary Gow and Kitty Werner! Almost immediately drafts of the first chapters of the book were showing up in my email inbox! It is truly remarkable how both of them seemed to absorb the multitude of personal experiences that constitute the history of Ward Lumber Company. As they peered into the lives of many generations of the Ward Family, they discovered the lives of an entire town and its inhabitants intermingled as they tried to survive and live gracefully during the early years of Vermont, back in a time when comfort and prosperity were elusive and reluctant companions. Thank you Mary and Kitty.

We also want to thank everyone who gave support and assistance to Mary and Kitty, including those who offered their time for interviews and reflections!

Holly Ward was my father's cousin and partner in business, and he was our partner in this historical endeavor. We thank him for all of his invaluable time and assistance in digging up old photos and the hours spent with Mary and Kitty recalling the past, while working in tandem with my father to explain the inner workings of the mill and its operations. I know that Holly takes great pride in keeping the small clapboard mill operating to this day. It is the only remaining vestige of Ward Lumber Company that remains since its glory days when large robust mills hummed with activity up and down the valley. By keeping this small specialized and historic mill operating, Holly has extended the history of Ward Lumber Company to 140 years!

This book offers a small window into the lives of some of Vermont's early residents and the struggles and perseverance that it took to survive. It is also an interesting portrayal of the challenges that pioneers of industry in America faced when this country was young and aspiring to become the remarkable country that it is today. I want to thank everyone involved!

Whitney Ward
Naples, Florida
January 16, 2012

TABLE OF CONTENTS

In the Beginning…	1
Ward Family History to 1874	5
Hiram O. Ward, the early years	7
Hiram and his Mills	12
Ward Lumber — 1889 to 1900	17
Burton and Clinton Ward, the Exposition and the Big Crash — 1893	23
The Ward Big House — 1901	25
1914–1917 Time of Sadness	31
The Turbulent Years 1925–1940	35
Brighter Side of the 1930s	51
The Ward Stores	55
The Family Through the Years	61
Reforestation circa 1910 to 1968	73
Logging & Timberlands	83
Mills and Water Power	91
Surveys and Changing Land Use	107
Move to Waterbury	109
Merlin's Speech	111
1968 — Sold	115
Owen Miles Ward	117
Holly Merlin Ward	119
The Fourth Generation Siblings	121
The Clapboard Mill	123
Ward Lumber Company Timeline	131
The Diary of Hiram O. Ward, 1874	133

Mr. & Mrs. Earl and Elizabeth (Munson) Ward
Hiram Ward's Parents

In the Beginning...

Monday, February 2, 1874

Thermometer 42 below zero. Go to Waterbury with lumber and get meal and flour &c. draw it to Dowsville & come home at night

Hiram O. Ward wrote this journal entry in his little diary, his script noting the day's events in pencil. Frigid weather barely slowed his relentless pace. That week, he went to the mill and "worked at the waterwheel some," he drew logs, took a hog to Moretown. For part of several days he split wood. Friday was again notably cold; he adapted and spent that day whitewashing the kitchen and splitting more wood. Saturday he was back in Dowsville drawing more logs. On the Sabbath, the meeting at church was cancelled and Hiram Ward remained, "home all day."

In 1874, the year of this surviving diary, Hiram Ward was 34 years old. Enterprising and unflinchingly hard-working, he was active in his Duxbury, Vermont, community and church. While his farm on Ward Hill was productive, he was beginning to build his lumber business. With his brother William Ward, he had recently purchased a sawmill on Dowsville Brook. In a few years time, Hiram Ward was celebrated as "the leading businessman of the town," in *The Gazetteer of Washington County 1783–1889*.

The lumber business, as started by Hiram Ward, thrived for almost a century. An industry leader in this region, Ward Lumber Company earned the distinction of being one of the state's oldest lumber mills in continuous family operation. Four generations of the family served at the helm of the business: Hiram Ward, widely known as H.O.; his son Burton Smith Ward; Burton's sons

Kenneth Hiram Ward and Merlin Burton Ward; their sons Owen Miles Ward and Holly Merlin Ward. Based in Moretown from the late 1800s, Ward Lumber Company was, by far, the town's foremost employer, providing scores of jobs and economic stability to many in the community.

The history of the Ward Lumber Company is an American success story of hard work and entrepreneurial spirit. Through the decades, the Ward family steered the business through changing and often challenging times. Hiram Ward was drawing logs and milling lumber when the Civil War was still a fresh memory; his family continued that work through two World Wars, the Great Depression, the booming post-war '50s, up to 1968 when the company with its vast landholdings was sold. Communication, technology, and transportation revolutionized over those years. Keeping competitive and serving new markets, the company changed too, innovating and adding efficiencies to its operations.

Challenges faced by the Wards were often daunting. Three times fires devastated the company's mills—two of the conflagrations were accidental, one was an arsonist's work. Historic floods damaged dams and tore hundreds of thousands of feet of lumber from the yards, carrying some as far as Lake Champlain. Working with crushingly heavy logs and powerful wood-slicing saws was always dangerous; not surprisingly, tragic accidents occasionally took a toll in injuries and even death.

Ward Lumber Company was in the business of transforming natural resources to usable products. Trees that grew in central Vermont forests were harvested and milled by the Wards and their employees. In many shapes and dimensions, this Vermont hardwood and softwood went to all corners of the country and beyond. Rocking chairs, parlor organs, Steinway pianos, thousands of cribs and playpens were among the vast array of household furnishings made from the company's lumber. Trains rolled on Ward Lumber railroad ties. Giant hemlocks from Ward land in Warren helped build the St. Lawrence Seaway. In the era before corrugated cardboard, wooden packing boxes held products from apples to lamps to wringer washing machines—the Wards made boxes, lots of boxes. From the shavings they made excelsior, spaghetti-like wood strands

Photograph courtesy of the Duxbury Historical Society

Dowsville Church, as it appeared in the 1800s above, and as it appears today. The little building to the far left is the outhouse that is still in place. This little church played quite a part in the Ward family history, as Hiram was one of its biggest supporters.

that pre-dated Styrofoam peanuts as packing material. Early in his career, Hiram Ward cut clapboards; through the decades, clapboards were a Ward mainstay. Dimension lumber milled by the Wards constructed countless buildings. In the 1950s, Ward mills cut and sold lumber sets of studs, rafters, siding, interior trim—everything needed for an entire house. These were trucked to Burlington and assembled as new homes in the growing city.

Ward Lumber Company's history is also a story of connections between human endeavor and the land. Hiram Ward, his son, grandsons, and great-grandsons, recognized the importance of restocking the forests. The family acquired massive land holdings, upward of 30,000 acres, to assure sources of lumber. From the early 1900s, the Wards managed planting of seedlings and pruning on their tree plantations. More than 1,500,000 trees from the Vermont State Tree Nursery alone were planted on Ward lands. Forestry students and academics from the University of Vermont, Yale, Syracuse and other universities visited and studied their plantations.

The Ward Lumber Company and the Ward family were prominent in the region and the state. For decades, the livelihoods of many in Moretown and beyond relied on the company—men worked in the mills, others were loggers felling and hauling trees. Together, the Wards and community travelled together through eras of phenomenal change.

Cutting clapboards at the sawmill in Moretown. The job is still done on the same equipment used in H.O.'s time.

Ward Family Genealogy

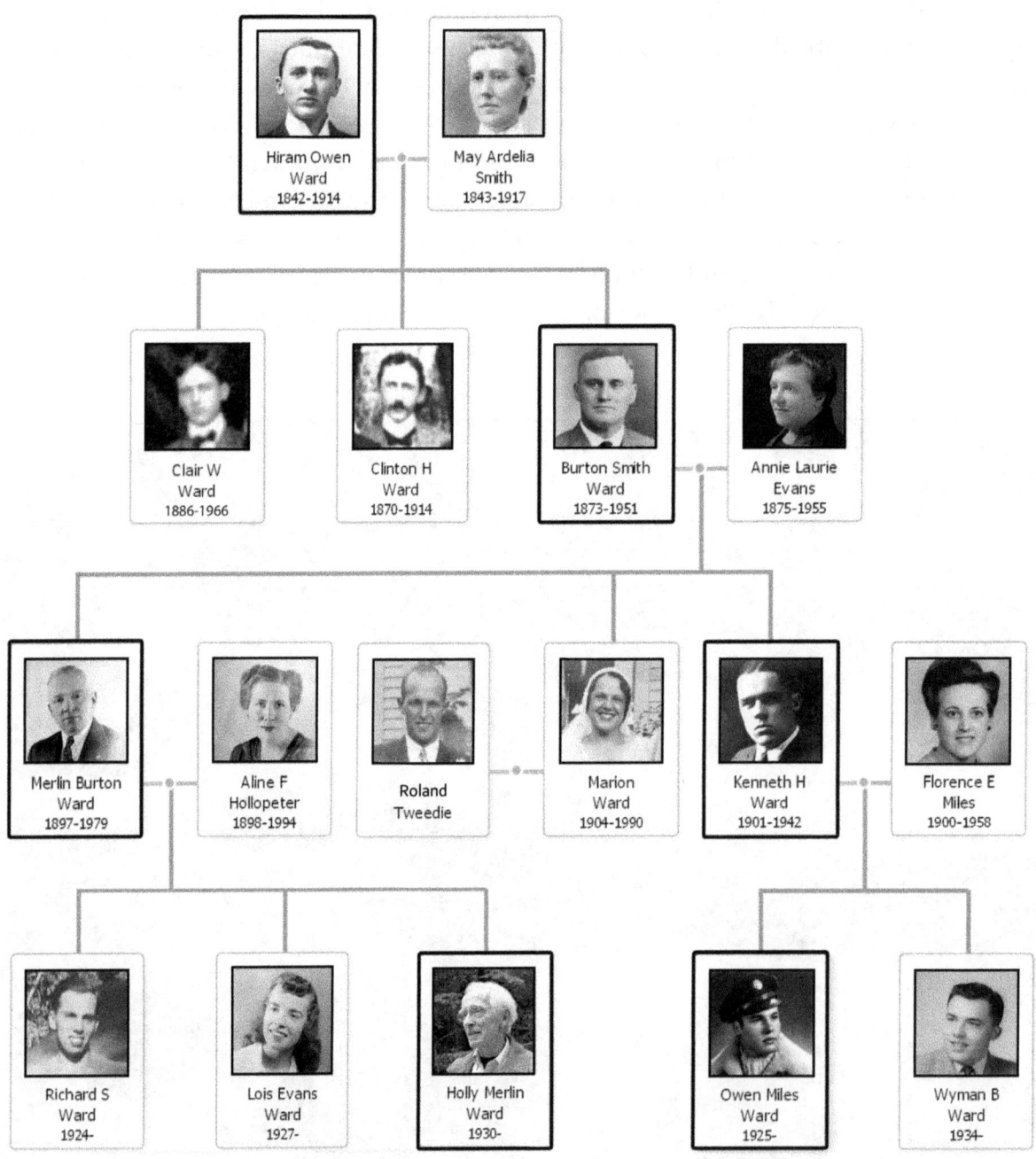

Ward Family History to 1874

The Ward family's lumber business history starts in 1872—the year that Hiram Ward and his brother William purchased a little water-powered sawmill at a cascade on the upper reaches of Dowsville Brook in Duxbury. The circumstances that brought the Wards to this endeavor stretch back far earlier.

Hiram Ward, born January 10, 1842, belonged to the eighth generation of descendants of William Ward, believed to be originally from Derbyshire, England. An early Massachusetts Bay Colony settler, William Ward was one of the founding settlers of Sudbury, Massachusetts. Sudbury was the 19th permanent town in the colony, and only the third inland town. An early Sudbury town map shows that William Ward had twenty acres in the northwest part of town, on the road to Concord. William Ward and Elizabeth Eastman, his second wife, apparently came to the colony together from England. In all, this William Ward had fourteen children; the eighth was his namesake.

This early history of the Ward family is preserved in detail in an 1851 genealogy written by Andrew Henshaw Ward who lived from 1784 to 1864. The 265-page volume is titled *Ward Family; Descendants of William Ward, who settled in Sudbury, Mass., in 1639*.[1]

Hiram Ward's side of the clan spent five generations in Massachusetts before venturing into the young state of Vermont. Along the way, notable ancestors included Colonel William Ward, born March 27, 1680. This William Ward found considerable success surveying newly chartered towns; new townships were popping up in Massachusetts like spring flowers in

---

[1] The book is in some private and library collections, it is also available online in its entirety through the Open Library. http://www.archive.org/stream/wardfamilydescen00byuward#page/46/mode/2up

Cascade on Dowsville Brook on Ward Hill where the original H.O. Ward mill was located. The foundations are to the right.

those days. Beyond surveying, William Ward was a magistrate. In his work helping establish townships, he acquired extensive landholdings. His son, Hezekiah Ward and his namesake son, Hezekiah Ward continued this branch of the Ward family. They lived in Marlboro, then Southboro, and then Grafton, Massachusetts in the 1700s.

Artemus Ward, a Ward cousin of the 18th century, distinguished himself as a Revolutionary War leader and hero. Ward had served the Crown during the French and Indian War, but became a prominent spokesman against the British Acts of Parliament. The Massachusetts militia opposing the British elected him general and Commander-in-Chief. In June 1775, the Continental Congress appointed George Washington as Commander-in-Chief of the "Continental Forces raised or to be raised for the defense of American liberty." They appointed Artemus Ward as Washington's second-in-command. A letter written to Artemus Ward from George Washington in early 1776 attests to his role in the colonies' march to independence.

The third Hezekiah Ward broke with his Massachusetts roots and ventured north to Vermont. The circumstances of his move are not known, but on January 14, 1812, Hezekiah's wife, Jemima Johnson, died in Burlington. Vermont was a young state, having joined the Union barely twenty years before. Widowed, Hezekiah Ward married Ruth Stockwell, a Duxbury widow. Ruth pre-deceased him. His third marriage was to Elizabeth Eastman. Among Hezekiah's thirteen children was Earl, Hiram Ward's father. Hezekiah died in Duxbury on September 28, 1849 in his 79th year.

Earl Ward, born in 1800, married Elizabeth Munson of Duxbury on March 12, 1828. They had six children—the eldest, William Edward Ward was born in 1829, the youngest, Hiram Owen Ward was born January 10, 1842.

The graves of the original residents of Ward Hill, including Hezekiah and his second wife, Ruth, located in a small cemetery in Duxbury on the DeLong Road, as seen in the photo on page 9.

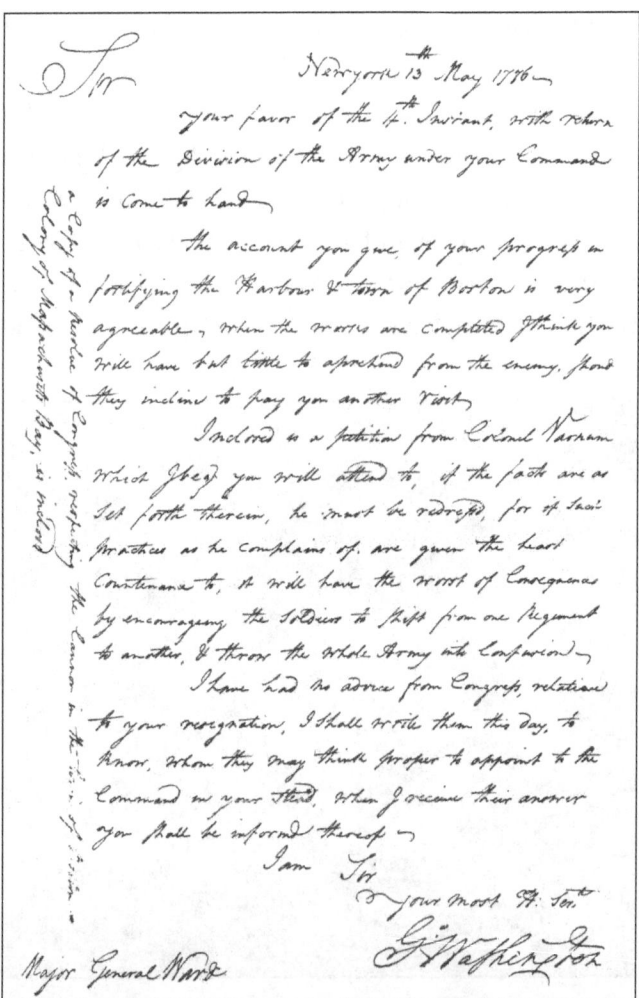

George Washington's letter to Artemis Ward

Hiram O. Ward
The Early Years

Change was in the Vermont air of Hiram Ward's childhood. The 1840s, when he was toddling around his family's Duxbury farm, were the years when the peak clearing of land for agriculture in Vermont was reached, according to the Vermont Division of Forestry. The early clearing of Vermont's native forests was done primarily to open land for farms. Much of the timber of the native forests was burned—some simply for clearing, some in limekilns and for the production of potash. Early agriculture in Vermont got an unexpected boost when Merino sheep were first imported into the state in 1811. These prolific wool-producers thrived here—they were hearty enough to endure the seasons and they grazed successfully even on the rocky pastures of hill farms. The sheep boom was short-lived. In 1837, there were over 1,000,000 Merinos in Vermont—about four times the human population. Changes in tariffs and the opening of fertile flat land in the Midwest curtailed the sheep surge. During Hiram Ward's childhood, the ovine numbers were already dwindling as raising sheep became less profitable. Sheep pastures were among the lands that would later be abandoned in the state—and farm abandonment tied into Ward Lumber Company's reforestation efforts.

From the colonial era, sawmills figured prominently in the Green Mountains. The earliest documented Vermont sawmill was in Westminster in around 1738. Simple water-powered sawmills were terrific labor savers for settlers, and they proliferated. The 1840 census reported 1081 sawmills in the state. These little operations primarily served the building needs of their immediate communities.

A major engine of change was entering Vermont during Hiram Ward's childhood years. In 1846, ground was broken for the Central Vermont

Railroad, the state's first. With its main line running through White River Junction, Middlesex, and Waterbury to Burlington, it opened access to a new world of markets in Boston, New York, and beyond. Trains were running on the tracks in 1849. In Duxbury, the railroad was undoubtedly big news.

Eastman Business College Courtesy The Early Office Museum

Eastman Business College

The Eastman Business College, founded in 1859, had a hands-on rather than theoretical approach. Students learned by doing. Eastman even printed its own currency for its banking and finance classes. Eastman's founder was a cousin to George Eastman who later founded Eastman-Kodak Camera Company. S.S. Kresge, founder of the Kresge stores (now Kmart) was another Eastman alumnus.

The youngest of his family, Hiram Ward studied at his local district school in Duxbury. As a teen he went to the Barre Academy. His academics would have required him to board in Barre—commuting 25 miles was not possible. As a young man, Hiram went out of state to continue his education at the Eastman Business College in Poughkeepsie, New York.

The 1860s were a busy decade for Hiram Ward. Part of the time he was working on the family farm in Duxbury, as noted on the 1860 U.S. Census. He registered for the military draft in 1863 during the Civil War, but apparently did not serve. He studied at Eastman—possibly toward the middle of the decade. In 1866 he married May Ardelia Smith of Hopkinton, New York. May, also sometimes called Marion or Mary, had been living with her aunt and uncle, the Canerdys, in Duxbury when they met. Hiram and Mary had three sons: Clinton, born in 1870; Burton Smith born February 13, 1873; and Clair born in 1884.

Hiram and May Ward

Photo by G. H. Dale, from the collection of Skip Flanders

The covered bridge crosses Dowsville Brook in South Duxbury on what is now Route 100. In the background is the steeple of the South Duxbury Church on the left of the road in the distance. The original cemetery on what is now DeLong Road is on the right. The photo was taken in 1900–1910, south of what is now the Harwood Union High School.

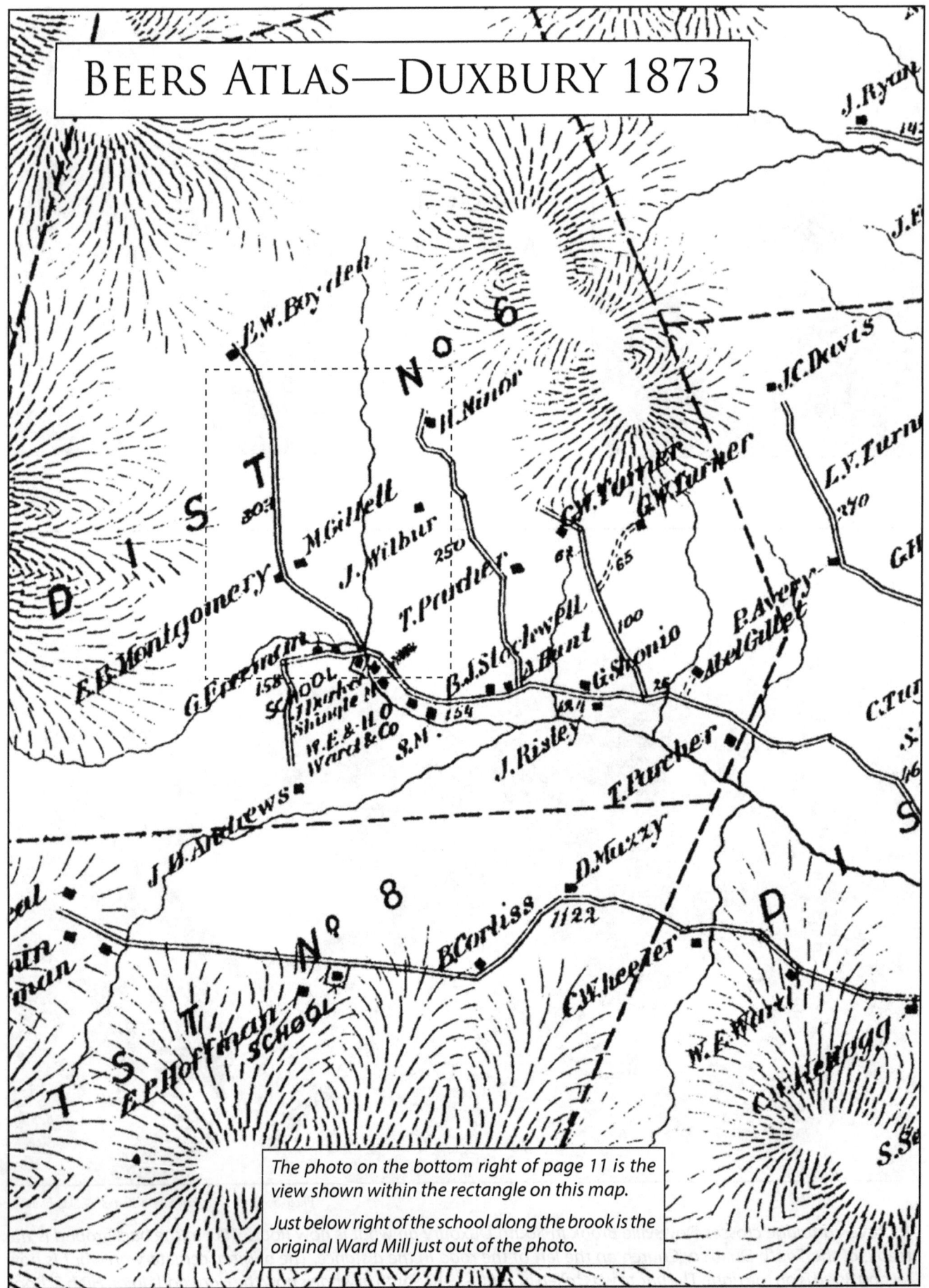

Beers Atlas—Duxbury 1873

The photo on the bottom right of page 11 is the view shown within the rectangle on this map.

Just below right of the school along the brook is the original Ward Mill just out of the photo.

Ward Hill Late 1800s

Two farmhouses on Ward Hill

Daniel Murray farm on Ward Hill 1858 mis-identified on the map opposite as D. Muzzy, on the lower road

Photographs on this page courtesy of the Duxbury Historical Society

Dowsville at the time of the Wards near the top of Ward Hill Road. The Dowsville School is the lighter building right-center. Vigilante Road is above in the background. The Freeman home and barn are the buildings on the left center on the hill. The map on page 10 shows the location.

Hiram and his Mills

A typical mill of the 1800s built along a small stream. While this isn't known to be one of the Ward Lumber Company mills, it is typical of the era.

A glance at any of the beautifully detailed *Beers Atlas* maps of the 1870s (a section of Dowsville, page 10) shows the proliferation of mills along state waterways in those days—sawmills, shingle mills, box mills, grist mills and more. The mill that Hiram O. Ward and his brother William purchased in 1872 was one of these.

With abundant streams and rivers running down mountains and hillsides, the potential for water power was immense in Vermont, as early settlers recognized. The trick was to harness the power of water to do mechanical work. A mill was usually sited at a naturally occurring narrow place in a stream or river with a drop in elevation. Typically, a simple dam built of timbers was installed in the waterway. Water was funneled from the pool behind the dam through an artificial course called a millrace. The water tumbling down the millrace turned a waterwheel. In the early most simply designed sawmills, one end of a long piece of wood, the "pitman arm" was attached to the wheel—not at the hub but near the perimeter. The other end of the "pitman arm" was attached to a wooden frame that looked much like a window frame. A saw blade was secured vertically in the frame. As the wheel turned, the pitman arm moved the frame and saw up and down. The saw's sharp teeth cut the wood on the down stroke. The sash was also geared to a "carriage," a platform holding the log. On the upstroke, the sash would pull the carriage forward, moving the log to the saw to be cut. After a log passed through the saw, the miller pulled the carriage back for another. Hiram Ward's first mill on Dowsville Brook probably followed this popular labor-saving design.

Straight saw, circa 1830s

From the beginning, Hiram Ward brought his business sense to his lumber enterprise. Soon after buying the mill, he invested in a circular saw, a fact noteworthy enough to be mentioned in the Duxbury section of the *Gazetteer of Washington County Vermont 1783–1889*. Circular saws were fairly recent innovations, and not widely used until the latter part of the 19th century. The circular saw had a couple of major advantages over the old sash saws. It cut a much thinner kerf through the wood, wasting less lumber than a straight saw. It was also fast—really fast. A smoothly running circular saw in that era could cut as much as 1,000 to 2,000 feet in an hour. The early circular saws had quirks and required different maintenance than their simpler counterparts, but their advantages led to a new age of efficiency. Lane Manufacturing of Montpelier, a major manufacturer of mill machinery, got a patent in the early 1870s for a very successful circular saw. Over the years, Ward Lumber Company owned many pieces of machinery made by Lane. This may have been one of the Wards' first Lane saws.

The lumber business clearly agreed with Hiram Ward. In early 1875, he bought his brother's share of the mill and its land. In short order he was diversifying his business and producing a range of products from several mills. In the 1880s he was producing packing boxes, dimension lumber, clapboards and even organ boxes. Parlor organs, also called reed organs, were hugely popular in that day and were a beloved fixture in small churches and in homes.

Beyond growing his business in Duxbury, Hiram Ward served his community. He held the position of Town Superintendent for several years and was elected Selectman in at least 1888 and 1889. In both 1886 and 1888 he served as Duxbury's Representative in the Vermont Legislature. Hiram Ward's lengthy obituary written decades after his early Duxbury days still expresses that town's respect for and fondness of him.

For Hiram Ward and others in the lumber business, the 1880s was an extraordinary decade. Vermont's all-time peak saw-lumber production was in 1889—375,000,000 board feet of lumber were cut in the state. Thirty-three years before, in 1856, that total was a mere 20,000,000. Hiram Ward was among the businessmen who led that growth. He recognized and seized opportunity,

Left: The remains of the Dowsville Mill along the Dowsville Brook. Abandoned generations ago, all that remains are these foundation stones of the mill itself.
Right: the remains of the foundation of the Dowsville Mill office building uphill from the mill, along the Dowsville Brook.

he invested in his industry, and he reached out to growing markets.

The Duxbury entry in the *Gazetteer of Washington County 1783–1899* noted:

Hiram O. Ward, son of Earl and Elizabeth (Munson) Ward, was born in Duxbury, January 10, 1842. He received his education in the district schools and Barre Academy, and graduated at Eastman Business College. About 1866 he married Mary A. Smith, of Hopkinton, N.Y., and they are the parents of three sons. Mr. Ward is now settled on road 27, and is the leading businessman of the town. He is an extensive dealer in and manufacturer of lumber, owns several saw-mills in Duxbury and other towns, manufactures packing boxes, and deals in pianos and organs. He is deservedly popular with his townsmen, is now chairman of the board of selectmen, and represented Duxbury in the legislature of 1886 and 1888. [p. 254]

Hiram Ward's investment in Moretown shows his business acumen. After starting his lumber business on Dowsville Brook, he expanded elsewhere in Duxbury with further mills, and also dipped into Moretown. Moretown village had grown up around a spectacular natural cascade on the Mad River well-suited to development of water power. Moretown townsfolk recognized the potential; they also knew that investment was needed to develop industry there. *The History of Washington County* in the *Vermont Historical Gazetteer*, collated and published by Abby Maria Hemenway in 1882, includes a wistful Moretown wish.

The Mad river affords some of the best water privileges found in the State, and should the inhabitants of Moretown induce some moneyed firm to put in a large manufacturing house here, thus utilizing more of the water power, and urge the building of a contemplated railroad, which has already been surveyed through the town, it would greatly develop the resources of and build up our town. [p. 597]

Hiram Ward was the man to fill the order. According to local newspaper reports of the time, H.O. was already leasing part of the old sash and blind factory on the Mad River in Moretown from the mid-1880s. After the mill burned in 1887, he purchased the site and built a new lumber and grist mill there. This site would be expanded and improved through decades as the Ward Lower Mill.

Moretown village was already home to three others, and more mills with other owners were elsewhere in town, including on the Winooski and on Jones Brook. In the village, Messrs. Parker

Moretown.

One of the saddest events ever recorded in the history of our town occurred last Sunday morning, about one o'clock. The mill known as the old sash and blind factory, and occupied by H. O. Ward as a box factory and grist-mill, burned to the ground and J. B. Fassett and wife, who occupied rooms over the mill, perished in the flames. Earlier in the evening a fire was discovered in a hanger supporting a shaft near some gearing by Mr. Dale (who was running the grist-mill) and Mr. Fassett. They extinguished it, as they thought, and nothing more was seen of it until the mill was discovered by the neighbors to be in flames, and it was then too late to save anything in the building, or even the occupants, from whom no outcry was heard. There was so much combustible matter in the building that it burned very quickly. After the walls fell in, the body of Mrs. Fassett was discovered outside of the wall so burned as to be unrecognizable, but the small part unconsumed of the body of Mr. Fassett was not found until afternoon Sunday. No alarm of fire was raised, so that many of the villagers knew nothing about it until morning. Besides the machinery, etc., in the mill there was quite an amount of grain owned by customers and between fifty and seventy-five thousand feet of lumber owned by Mr. Ward. The machinery and stock were partially insured. Our village has been unusually free from loss by fire, and this event called together a large number of people to view the ruins. ==== Quarterly meeting services will be held here Sunday, May 29. Presiding Elder Truax will be present.

The Vermont Watchman *May 18, 1887*

Courtesty UVM Special Collections

Lower Mill pond, circa 1907–1915

The pond created by the lower mill's dam.

and Gillett turned out 200,000 to 300,000 feet of lumber and 200,000 shingles annually. Robinson's grist and sawmills cut a little less lumber and made about the same number of shingles. The Lovejoy and Towle carriage and undertaking shops handled undertaking and also made and repaired carriages, wagons, and sleighs. The village was also home to two general stores, a hardware store, an hotel, two blacksmith shops, a Roman Catholic church, and a Methodist church. About 200 people lived in the village.

The 1889 *Gazetteer* also noted:

> H.O. Ward's box factory and grist mill, at Moretown village, on Mad River, gives employment to six men, and manufactures about 1,000,000 feet of lumber into boxes annually. An old mill was burned on this site May 15, 1887, and J. B. Farrell and his wife who lived in one part of it, perished in the flames. Mr. Ward's mill was built in the ensuing fall. Charles H. Dale operates the grist mill and grinds from 20,000 to 25,000 bushels of grain each year. Mr. Ward resides in Duxbury. [p.402]

The following year, he, May and their sons moved to Moretown. A new era in the Wards' lives and the history of Moretown opened in 1890.

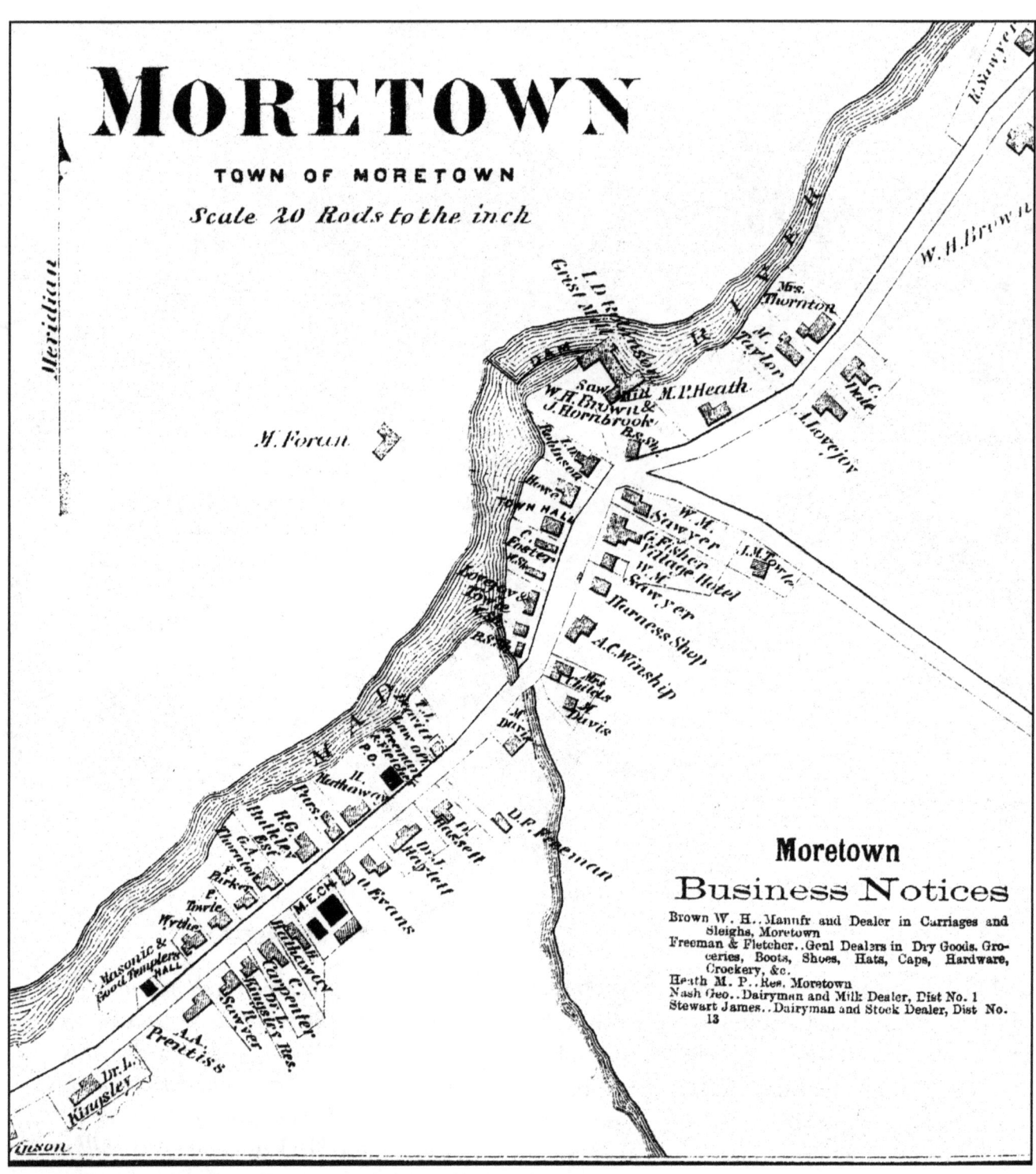

1883 Beers Atlas
Business Notices
Moretown

Note the location of the grist mill at the river bend, eventually purchased by H.O. Ward

Ward Lumber — 1889 to 1900

In September 1889, H.O. Ward and his family moved from Duxbury to Moretown. Hiram and May's sons were growing up: Clinton had just turned 19. Burton Smith Ward was 16. Clair was 5 years old.

Articles and news columns from *The Vermont Watchman* and *State Journal*, a weekly newspaper published in Montpelier during the late 1800s and early 1900s, provides many glimpses of Hiram Ward and his thriving business during the last years of the 19th century. In every issue, *The Vermont Watchman* included "town sections," short columns with local news from different communities. With Hiram Ward's prominence as a businessman, employer, church member, town officer, and county senator, he was mentioned often in this newspaper. More than a century after they were written, reports from *The Vermont Watchman* offer a look at H.O. Ward as he was known in his time. They also provide a wonderful sense of Vermont community feeling of that era.

"H.O. Ward moved into town the past week from Duxbury, moving into the house lately purchased by him of Philo G. Towle," a September 1889 issue of *The Vermont Watchman* reported as Moretown news. The column further noted, "H.O. Ward is walking on crutches, the result of a badly sprained ankle." Other Moretown news included, "Nearly everybody attended the Mad River Valley fair and a few went to the state fair at Burlington." At the Burlington fair a Mr. Herrick of Middlesex, "had his pocket picked of one hundred sixty dollars." Mad River Valley fair-goers

> **Waterbury Items.**
>
> Among the enterprising business men of this vicinity, quietly pursuing "the even tenor of his way," is Mr. H. O. Ward, manufacturer of lumber. His residence is on the Duxbury side and his mill is at Dowsville in that town, but his business is done in this village and his shipments are made from the railway station here. He owns about nine hundred acres of timber land in Duxbury, buys largely of the farmers in that vicinity and takes the product of mills at South Duxbury and in other localities. In the year 1882 Mr. Ward sold 810,000 feet of lumber and sawed 160,000 feet for customers, making his total transactions nearly a round million feet. For the last two years he has bought from 2,500 to 3,000 logs from his neighbors, in each year. Of his sales 300,000 feet was clapboards, 150,000 feet hard wood and the balance, spruce dimension stuff and boxing. Every year wood and lumber are becoming more scarce and more valuable articles. The forests on these rough hills of Duxbury, Waterbury and Moretown, if worked judiciously, with a wise regard to their preservation and the reproduction of their stores of wood and lumber may afford an income to the prudent lumbermen as permanent and relatively as profitable as the pastures and the tilled fields to the farmer. But in the management of these woodlands the principles of good forestry should be understood and regarded. For the most part these lands are rough and rocky, valueless except for growing timber. If they are "skinned," fire and the washings of the rains and melting snows will make an end of the light vegetable mould which to a great extent forms the soil, and if a new growth gets a foothold at all the chances are it will be wild cherry, poplar or white birch instead of the more valuable kinds which compose the original growth.....

From The Vermont Watchman and State Journal *(Montpelier, VT) April 18, 1883*

had a better time, including hearing music by the Gouldsville band. "Among other entertainments at the Mad River Valley Fair was a free fight in the evening of the second day by the venders of picture frames, peanuts, etc."

A March, 1891, issue of *The Vermont Watchman* announced, "H.O. Ward has put into his Newhall Mill a 40-horsepower engine." The

Lowell, Massachusetts textile mills on the Merrimack River

Library of Congress

Ward Newhall mill was on the Winooski River. Three months later, a June, 1891, issue noted the impressive productivity of Ward mills, thanks in part to the new engine.

> H.O. Ward has shipped from Middlesex station this last week, in round numbers, two hundred twenty-two thousand feet of lumber. He has three mills, two of them running night and day, up to their full capacity, with every man that can work to advantage. He also has nine two-horse teams on the road delivering his lumber to the depot and is two weeks behind his orders. He manufactured nearly every foot of lumber last week that he shipped.

Within a few years of investing in Moretown, Hiram Ward was the town's leading employer. Thirty men were on his mill payroll, crews of loggers cut and hauled logs in the winter, and he was shipping many railroad carloads of lumber and boxes from Middlesex each month. An August, 1893, article conveys some of the local enthusiasm about his success. J.W. Bates was Hiram's right-hand man and is remembered for his enormous contribution to the success of the company.

Ward's Shops and Mills

The name of H.O. Ward has long been identified with the business interests of Moretown and that part of Washington county drained by the Mad river. Mr. Ward was in Massachusetts last week, but J.W. Bates, his foreman, courteously devoted his time while the newspaper man was there to showing him over the mills, and submitting good-naturedly to any amount of pumping. In addition to the large business done by Mr. Ward at his own saw-mills, clapboard mills, grist mills and box shop, he has contracted for the year's stock of coarse lumber and clapboards that three large mills in Fayston can produce, and besides is obliged to buy considerable outside to fill his orders. The product of all his mills is loaded at Middlesex, seven miles distant, an average of eighteen carloads a month the year through. Mr. Ward also deals in flour and feed of all kinds, doing business of $60,000 a year. He has thirty men on his payrolls, and during the winter when stocking up, employs a small army of choppers and teamsters. Mr. Ward ships his lumber to all the New England states, but principally to Massachusetts and New Hampshire. He owns large tracts of wood land in Dowsville and on Moretown mountain, enough to supply his mills for many years to come.

Through his sound leadership, management, and investment, Hiram Ward was running a exceedingly successful business. He was consistently included in the Moretown listings of the *Vermont Business Directory* for his "clapboards and dimension stuff, packing boxes, flour and feed." In Moretown village, he had both the Upper and Lower Mills on the Mad River. As of 1893, *Vermont Business* directories of the time no longer list him operating a mill in Duxbury.

A quick look down the railroad tracks can shed some light on the demand for lumber that helped fuel the growth of Hiram Ward's business. As lumber mills such as his were built on rivers to utilize the force of falling water to power saws, other types of mills around New England were also built on waterways to utilize that power. Francis Cabot Lowell introduced the power loom to Massachusetts in the 1800s, igniting a new river-driven textile industry. After the Civil War, with fresh availability of southern cotton and innovative new equipment, New England textile manufacturing took off at a frenzied pace.

Massachusetts was a leader in industrial innovation, but other northeastern states were bustling, too. The early Massachusetts and New England textile mills, like their cousins in Britain, used direct-drive hydropower. Turning waterwheels drove the gears and pulleys that drove the looms, spinning, carding, and sewing machines. In the 1870s, mill owners began supplementing that direct water power with steam engines. In the 1880s and 1890s, progressive mill owners turned to hydro-electricity. With this innovation, they still used water power, but instead of turning a conventional waterwheel, the water turned a turbine and generated electricity. The turbine was smaller and more efficient than the old waterwheel; a mill with hydro-electricity could do more work than its earlier waterwheel counterpart.

As bobbins whirred, looms clanked, and belts spun, New England mills attracted thousands upon thousands of workers. Workers from rural areas and immigrants from other nations flocked to cities in Massachusetts and New Hampshire. Closer to home, Winooski, Vermont, was another textile hub. As people were attracted by mill jobs, populations soared. The population of Massachusetts nearly doubled between 1870 and 1900 leaping from 1,457,351 people to 2,805,356. The state picked up another 500,000 residents by 1910. With this surge of workers and families came colossal demand for housing.

Middlesex landscape, July 3–5, 1912. Notice the train in the background. Ward Lumber Company logs rode those trains to points around the country.

In New England, one style of affordable housing was especially widely developed—the "triple decker" tenement. Three stories tall, with an identical floor plan on each level, a triple decker had a narrow façade but greater depth—many could be lined up on a new city block. These multi-family residences were made of wood construction and almost always had clapboard siding. Mill owners often built tenements and leased the apartments to their workers; some

With the textile mills' success, there was great demand for housing. This four-apartment tenement was home to four families working for the Chace Cotton Mill in Burlington, Vermont.

independent developers also built them. In Fall River, Massachusetts, for example, mill owners constructed 12,000 units of company housing in the late 1800s.

Beyond the tenements, there was also a building boom of grander homes for the entrepreneurs, managers, and executives of the age. Tens of thousands of Victorian, Queen Anne style, and Colonial Revival houses shaped new residential neighborhoods. In New England, almost all houses, big and small were built with wood, only a small percent were brick or stone.

Ward Lumber Company records from this era are long gone, but Hiram Ward's success and business trips to Massachusetts, such as the one mentioned in 1893, indicate that trees cut from central Vermont forests, milled by Ward Lumber, and shipped by rail, helped build the growing cities of this industrial era. At the same time two other Ward products were also in demand—flour from the grist mill, and boxes. As cardboard boxes were still in the future, wooden boxes were used for shipping all kinds of products.

A man of his era, Hiram Ward was innovative. A January, 1894, issue of *The Vermont Watchman* reported that, "He has recently had a telephone put into his office, making it much more convenient with his growing business." The same piece mentions, "H.O. Ward averages to sell from two to three carloads of flour and feed each week."

Along with his business acumen, Hiram Ward also was a leader in civic life. He had served two terms as Duxbury's representative to the state legislature. In 1896, he served as Washington County Senator. He was a board member of the Waterbury Savings Bank and Trust Company. Through the years he was an active and devout member of his church in Moretown, but he also always maintained ties to the South Duxbury church.

An August, 1899, issue of *The Montpelier Watchman* offers a final 19th century glimpse of Hiram Ward and the Mad River valley of that time.

The Drought and the Crops

With the rest of Vermont, Mad River valley is suffering from the drouth. Mad River is "mad" only in name, in fact it now has not the proportions of a respectable trout brook. Springs that have not been known to fail for the past forty-five years are perfectly dry, and with the exception of one mill at Moretown owned by Hon. H.O. Ward, barely a wheel is turning from Middlesex to Warren. Mill owners have discharged or laid off their men, and are waiting for rain, and in many pastures there is no water for stock.

Notwithstanding this, the valley from Moretown to Warren has an appearance of June freshness. Corn is looking finely, heavy crops of rowen are being cut, and the meadows wear a dark green hue suggestive of the month of roses: This is particularly noticeable from Moretown to Waitsfield. A dozen or more inquiries by the reporter as to this brought no satisfactory answer. To be sure, the soil is deep and fertile, neither sandy nor largely of clay, but that hardly accounts for the fresh appearance of these meadow lands while the rest of creation in this part of new England is parched, brown and withered by drouth of unprecedented duration.

On the river road from Moretown to Warren is 500 acres of ripening corn, with about an equal acreage of oats. Not an acre of oats has proved a failure and not a field of corn but what gives promise of abundant harvest. The hay crop was nearly if not quite up to the average as to quality and quantity, and potatoes are yielding generously. Small wonder it is that the prosperous and well to do farmers in this section have no desire to change their location or their occupation.

All in all, with his mills running, local agriculture thriving, and the serene natural beauty of the region, it sounds like the Mad River Valley of Hiram Ward was a remarkable place to live.

Overlooking the Mad River Valley, circa 1900, looking northeast from the top of Mt. Ellen. The property was owned by the Wards and sold to become Glen Ellen Ski Area in 1962. At the time, Whitney Ward remembers his father, Owen, commenting that he couldn't believe they would pay $35/acre for land "too steep to timber"!

Courtesy of the Moretown Historical Society

G.H. Dale photo of the old upper mill below the dam. Note the children on the rocks.

Burton and Clinton Ward
the Exposition and the Big Crash — 1893

By Brooklyn Museum [Public domain], via Wikimedia Commons
The Ferris Wheel at the Chicago Exposition of 1893.

The Ferris Wheel at the Columbian Exposition in Chicago in 1893 was American innovation's spectacular response to the Eiffel Tower. Gustav Eiffel's steel girder tower was the landmark of the 1889 Paris World's Fair. George Washington Ferris's response was monumental, too, and it moved. This first-ever Ferris Wheel, 264 feet tall, swept passengers up and around in 36 comfortable gondolas, each one seating 60 people. On their nine-minute ride, priced at 50 cents per person, they gazed down on the reflecting pools and over 200 buildings of the Exposition, also known as the 1893 Chicago World's Fair.

This massive fair was the most exciting event of the year. Over 27 million people attended it—approximately one-quarter of the United States population of the time. With cultural, educational, artistic, and entertaining events and displays, its spirit was of enthusiastic optimism for American industry. Commemorating the 400th anniversary of Christopher Columbus's New World expedition, the fair was open for six months, May through October, 1893.

The central plaza of the fair was the Court of Honor, known as The White City. These grand buildings designed by the country's leading architects were almost all temporary, constructed just for the fair, but they were awe-inspiring in size and style. The White City inspired L. Frank Baum's *Emerald City of Oz* and Walt Disney's Magic Kingdom. (Disney's father worked on construction at the Chicago Fair.)

With over 65,000 exhibits from countries, companies, states, and organizations, the fair covered a lot of ground. Almost all states built pavilions and showed off their local culture and regional products, often in entertaining and quirky ways. California's contribution included a 125-year-old palm tree, a medieval knight made of prunes, and a red wine fountain in its Spanish style building. Pennsylvania sent the Liberty Bell and President John Quincy Adams' baby clothes. Vermont's pavilion was designed like a Roman villa from Pompeii. Other countries sent treasures of all kinds: Johann Sebastian Bach's clavichord; the furnishings of a Bavarian Palace; ostriches from the Cape Colony, now South Africa. Canada's 22,000 pound "Giant Cheese" was a crowd-pleaser.

Industrial progress and innovation held center stage. Near the entry, the fair's impressive power plant awed crowds with its massive steam engines and dynamos. One of the most exciting pavilions was the Electricity Building, which included a model home with countless wonders of a new era of electrical appliances—electric irons, clothes washers, lamps, burglar alarms, even elevators. In a bidding war between Westinghouse and General Electric to electrify the fair, the cost efficiency of Alternating Current won out, shaping the future of the United States electrical grid. American companies showed off scores of new products, including Quaker Oats®, Cream of Wheat®, and Juicy Fruit® gum. In the Zoopraxographical Hall, visitors watched early "moving pictures."

For Clinton, then 23, and Burton Smith Ward, 20, it must have been thrilling to start out for Chicago in mid-October to attend the Exposition in its final days. To two young men from rural Vermont whose father had founded a successful and growing business, the visit to the exposition would be exciting while giving them a broader view of business and commerce around the country.

Long-distance travel was by railroad in the 1890s. Massive steam engines pulled passenger and sleeping compartment cars made of wood and steel, lighted with lanterns. Chicago was a great railroad hub, and with the Fair's stupendous success, rail companies had added extra trains to help those millions of fair-goers get there and return home.

In the early morning hours of Friday, October 20, Burton and Clinton Ward were on board the Number Nine Pacific Express, a westbound train headed into Battle Creek, Michigan. An eastbound Raymond and Whitman Special train carrying World's Fair passengers back to Boston and New York was on the same tracks. The trains were expected to meet at Nichols, Michigan. The eastbound train had been directed to pull onto a siding there to let the westbound train pass. The westbound Pacific Express was running late. The crew of the eastbound Raymond and Whitman Special failed to follow orders. At 3:45 in the morning, the crews of the two trains saw each other; both engineers reversed their engines, threw on the brakes, then jumped to safety. The trains collided seconds later.

Newspapers near and far reported the tragic event, one of the worst train wrecks ever…."When the collision took place the second and third day coaches of the No. 9 train going west were completely telescoped. A horrible sacrifice took place, the second coach cutting through the third coach like a knife, the roof crashing over the heads of the sleeping, ill-fated passengers in the third coach, completely entombing them in a fiery furnace," reported a Colorado newspaper. Twenty-six people on the Number 9 Pacific Express were killed, many more were injured. There were no fatalities on the Special, which reportedly had sleeper cars and was moving slowly.

Around the country, through Associated Press and other accounts, newspapers also reported Clinton and Burton Ward's heroism in helping passengers escape the burning wreck.

The Vermont Watchman reported:

> C.H. and B.S. Ward, sons of H.O. Ward of this place, had a narrow escape from death on the ill-fated train at Battle Creek, Mich., on their way to the World's Fair at Chicago. The *Boston Evening Record*, in its description of the wreck, says: "C.H. Ward jumped out of the left side of the car through a window, the seat in front of him being occupied by a woman and her baby. He managed to pull the woman through the window. She begged him to save her baby, but the little one was fastened beneath the seats and burned to death. The poor mother was nearly crazed and had to be removed by force. B.S. Ward came out through the roof in some manner, just how he cannot tell. He jumped off the roof and assisted out the Smith family from Fort Plain, N.Y., consisting of father, mother and two children, all four being more or less seriously injured, but miraculously escaped, while all around them, passengers less fortunate, were either killed outright or fastened down in the debris and burned to death before help could reach them." Mr. Ward received a telegram from his sons assuring him they were all right and had gone on to the fair.

The Ward Big House — 1901

As the 20th century opened, Hiram Ward was prospering. His mills were productive, his sons were adults, his first grandson, Merlin, was three years old. The Ward family had been living in Moretown for a decade. According to family lore, Mrs. Hiram Ward put forth a gentle ultimatum: "If you want to stay married to Mary, you'll build a new house." He complied. In 1901, the family moved into a gracious new home.

The Big House, the Ward's handsome Victorian-style home, three stories high with ample porches and lots of rooms, became a Main Street landmark. A delightful family photograph, dated August 15, 1901 shows the extended Ward family, forty-one strong, at the housewarming. Hiram and May Ward (also known as Mary) are there, so are their three sons and Burton's wife, Annie Evans Ward. One of Hiram Ward's two grandchildren, who would later lead the company, is in the picture. Merlin Ward, three-and-a-half years old is tucked between cousins in the front row. Kenneth Ward, born in 1900, was still a baby. He was very possibly taking a nap on that distant summer day.

Through the years, the Ward Big House was not only a family home, it was a focal point of Moretown. The Wards entertained and hosted meetings of local groups and organizations. Behind the house long sheds housed equipment, stables, and served for wood storage. The field behind the house was also a lively place—in the 1930s the local semi-pro baseball team, supported by Burton Ward, played there, attracting huge crowds of spectators.

Settled into the Big House, life apparently ran smoothly for the family and company for the early years of the 1900s. Beyond the mills, Hiram Ward added one of the two Moretown general stores to his holdings in around 1905. A general store of the time carried a very wide range of products—grain, groceries, clothing, shoes, tools. The store acquisition may have tied into assuring a source of supplies for the lumber camps where the men would stay for extended periods in the winter months, cutting trees and skidding logs to supply the mills.

Not long after moving into the house, Clair, the youngest of Hiram and Mary's three sons went off to college—enrolling at Stanford University in California. With transportation of the time, California must have seemed as far away as Europe. But California clearly agreed with him, and except for a period of a few years after his father died, California would be his home.

THE FAMOUS HOUSEWARMING PARTY 1901–1902

Top Row: Mr. & Mrs. Turner, Harriet Gilbert, Alice Bisbee, Mrs. Crandall (Si.Sam Turner) Mr. Ed Turner, his mother, John Bisbee, Mr. Wyman, Levi Munson, Clair Ward, Herbert Ward, Samuel Turner, Mr. and Mrs. Currier, Irwin Joslin, Belle & John Griggs.

Next Row: Mrs. Sam Turner, Mrs. Bates, Mr. & Mrs. Syobell, Holly Gilbert, Ward Joslin, Turner child.

Left of Steps: George Holly Gilbert, Burt Ward, Burton Ward, Flora Gilbert, Jay Ward, Miss Crandall, Mr & Mrs. Harlan Munson, Mr. & Mrs. Earl Ward.

On Steps: Mrs. Hiram Ward, Mrs. Wyman, Mrs. Burton Ward, Josie Ward, Brenda Ward, Retta Gates, Mrs. Crandall, Joyce Bisbee, Wilifred Gilbert, Merlin Ward, Florence Ward, Ruth Joslin, Ethel Ward, Bertha Gilbert.

Right of Steps: Lena Ward, Hiram Ward, William Ward, Clinton Ward, Mr. & Mrs. Amis Munson, Frank Ward, Owen Gates, Marjorie Ward, Jessie Ward, Aunt Emily Parker, Aunt Millie Ward, Mr. Crandall,

Paul & Helen Syobell   (Father Lee Munson).

Moretown Vermont
Relatives at House Warming for
Ward's Big House - 1901 - 1902
Built by Hiram + May Ward

- 17 - Hiram Ward
- 8 - May Ward
- 3 - Burton Ward
- 10 - Annie Ward
- 24 - Merlin Ward
- 32 - Clair Ward
- 19 - Clinton Ward

1. Prof. Gilbert
2. Burt C. Ward
3. Burton Ward
4. Flora Gilbert
5. Jay Ward
6. Miss Crandall
7. Lizzie Munson
7. Harlan Munson
8. May Ward
9. Aunt Wyman
10. Annie Ward
11. Josephine Ward
12. Henrietta Gates
13. Brenda Ward
14. Stella Crandall
15. Joyce Bisbee
16. Lena Ward
18. Nell Ward
19. Clinton Ward
20. Mr. + Mrs. Leve Munson
21. Frank Ward
22. Carl Ward
23. Mary Ward
24. Merlin Ward
25. Hiram Ward
26. Aunt Emily Parker
27. Josie Ward
28. Owen Gates
29. Marjorie Ward
30. Mrs. Wyman
31. Samuel Gumer
32. Clair Ward
33. Bertha Gilbert
34. Sam Gumer
35. Leve Munson
36.
37. Mr. Curmer
38. Mrs. Curmer
39. Irwin Joslin
40. Belle Briggs
41. John Briggs

*May and grand-daughter Marion
circa 1905*
Courtesy of the Moretown Historical Society

From postcards of the era, courtesy of the Vermont Historical Society and UVM Special Collections

Before the Big House was built, circa late 1800s

Moretown
Before and After the Big House

Circa 1909 (the date of the postcard), after the Big House was built. (See arrow at right)

Courtesty UVM Special Collections

Courtesty UVM Special Collections

Moretown, summer and winter, as seen from a hill above town, showing the Big House with all its stables, workshops, and outbuildings.

1914–1917 Time of Sadness

After declining health, Hiram Ward died in May 1914. Tragedy struck two months later. In Morrisville for a cousin's funeral, Clinton Ward was killed in a car accident. Three years later, May Ward's death followed.

May 13, 1914

HIRAM OWEN WARD
Death on Saturday of One of Most Prominent Business Men of Central Vermont.

In the death of H. O. Ward of Moretown, the Mad River Valley loses her largest employer of labor and a man quietly interested in the church and the development of community life. His was a busy life, active almost to the last and filled with the spirit of progress. Although not well for some time, his condition did not appear as serious until about two weeks ago, and when at Battle Creek, Mich., he began to fail. His son and a physician were summoned, and, accompanied by his wife, his son, C. H. Ward of Moretown, and Dr. S. L. Goodrich of Waterbury, the return trip home was made, arriving the middle of last week. He gradually failed, however, and last Saturday evening at 8 o'clock Mr. Ward passed away, the cause of his death being stomach trouble.

Hiram Owen Ward was born in Duxbury January 10, 1842, the son of Earl and Elizabeth (Munson) Ward, the youngest of six children. His father, Deacon Ward, gave his name to the hill where their farm was located, Ward Hill being one of the localities of South Duxbury. He was also one of the strong men in the little church in that place, and the deceased early professed his Christian faith and joined in the church. Here he had always kept his membership, although he had been zealous in work of other churches. He was educated in the common schools of Duxbury, at the old Barre Academy and the Eastman Business College at Poughkeepsie, N.Y. In 1866 he married Miss May Smith, daughter of Mr. and Mrs. Harrison Smith of New York State.

During their acquaintance Miss Smith was visiting in this vicinity and making her home with her uncle, the late M. C. Canerdy, then of Duxbury. Mr. and Mrs. Ward lived for a number of years on Ward Hill, but later moved to a farm nearer Waterbury, Mr. Ward erecting the present house on the farm, the one now owned by Merton Johnson. While living here the family was very active In the work of the local Congregational Church, Mr. Ward's voice being a great help in the music of the church. Their home, too, was often the scene of Christmas trees and community gatherings, and all that was for the good of the locality. At this time the family thought much of building in this village, owning for a time the lot upon which the Methodist Episcopal

Church now stands. But because of his increasing lumber business in Moretown, they decided in 1890 to move to village and a few years ago built a substantial home there, which has been one of great hospitality. Mr. Ward had become a large lumber dealer having mills in Bolton, Northfield, Moretown, Fayston and Duxbury, and also owned a large number of farms and timber tracts. Strictly by his own ability, he became financially successful, which has meant a good deal to the town. But his life will also leave the deeper influence which comes from a strong Christian character, trained from childhood in the Bible, and delighting also in the hymns of the church. There was with him always a constant source of strength to himself and his family. In the Methodist church in Moretown, where he had worshiped for a number of years, his true worth had been felt.

Politically, Mr. Ward had been honored by Duxbury, Moretown and Washington county. He held many offices of responsibility in Duxbury, representing the town in the Legislatures of 1886 and 1888. He also represented Moretown in 1892 and was county Senator in 1896. In all these bodies he was very efficient and served on important committees.

Mr. Ward is survived by his widow, three sons, Clinton H. and Burton S. associated in the business in Moretown, and Clair, cashier of the New York Life office in Los Angeles, Cal. Four grandchildren also honor his memory, and one brother remains, Rev. Earl Ward of Meredith, N. H. the last survivor of "Deacon Ward family." The funeral was held from his late home Tuesday afternoon. People arrived early and the spacious house and porches could scarcely give room to the numbers who came to pay their tribute the deceased. One large room was nearly filled with employees, eight coming from Bolton. T. J. Ferris was in charge and Rev. A. A. Mandige, pastor of the Moretown Methodist Episcopal Church, officiated. A quartet composed of Mrs. Roland Griffith, Mr. and Mrs. Will Kingsbury and Ernest Kingsbury, accompanied by Mrs. Haylett, sang two of Mr. Ward's favorite hymns, "Jerusalem, My Happy Home" and "Saving Grace." Mr. Mandigo's remarks were 'very helpful and he expressed strongly the faith of the deceased.' The floral remembrances were beautiful. The bearers were all nephews. W. N. Ward, of Burlington, H. G. Ward of Moretown, Jay Ward of Meredith, N. H., Carl Ward of Hanover, N.H., Rev. O.H. Gates of Cambridge, Mass., and Rev. C.M. Gates of Wellesley Hills, Mass. During the winding of the procession to the little South Duxbury cemetery, where interment was made, a large number of employees walked just ahead, a distance of nearly two miles. Here, almost in the shadow of his old home on Ward Hill and, near the little church which he had always honored with his membership, the burial service was given by Rev. Mr. Mandigo, assisted by the Rev. O.H. and Rev. C.M. Gates. Besides large numbers from near-by towns, among those present at the service were Rev. Earl Ward and Jay Ward of Meredith N.H., Rev. and Mrs. Gates of Cambridge, Mass., Rev. C.M. Gates of Wellesley Hills, Mass., Carl C. Ward and Miss Lena Ward of Hanover, N.H., Mr. and Mrs. Ami Munson of Morrisville, Mr. and Mrs. George Crandall of Berlin, Mr. and Mrs. W.N. Ward and daughter Florence, and Mr. and Mrs. Guyett of Burlington, S.A. Allen, Mr. and Mrs. J.C. Briggs and M.H. McAllister of Barre, Dr. Mayo and Hon. Charles Plumley of Northfield, G.E. Moody and Dr. S. L. Goodrich, Judge E.W. Huntley. Mrs. E.F. Palmer, Jr., Mrs. Edward Cullen, Samuel Baird, and Mrs. Ramsdell from here.

EMF car 1912—Clinton driving H.O., with Burton (right back) and Clair as backseat passengers

CLINTON H. WARD KILLED.
Prominent Moretown Resident Meets a Fatal Auto Accident in Morrisville

Clinton H. Ward of Moretown, son of the late H.O. Ward, who was in Morrisville today with his mother and brother, Burton, and the latter's family to attend the funeral of his cousin, Harlan R. Munson, died at 7.30 this evening as the result of injuries received when his automobile was overturned as he was returning to the Munson home from the cemetery.

The accident occurred at 4.30 this afternoon, while he was rounding the turn near the C.H. Small residence. Running at a low rate of speed Mr. Ward in some unaccountable way lost control at his machine; which ran for a little distance on the edge of the bank, when it swerved to the left, the car going down the bank and turning completely over. Mr. Ward's breast was crushed against the steering wheel, causing internal hemorrhages resulting in death.

After the accident Mr. Ward was conscious and walked for a short distance, then he fainted. After being taken by ambulance to the Munson house he regained consciousness and remained in that condition until the last.

Mr. Ward was alone in the car, his mother, brother and family having returned from the cemetery by carriage. The body will be taken to his home in Moretown on Thursday, where the funeral service will be held. Mr. Ward was 40 years of age.

Ward Lumber Company with its mills, store, and vast landholdings had passed to the second generation. Abruptly, only two of the three members of that generation were left. Burton Ward and Clinton Ward had already been managing the business. Clair Ward, who had been living in California, returned to Vermont.

Three years later, on May 24, 1917, their mother, May Ward, died.

DEATH OF MRS. MAY S. WARD
Highly Respected and Long Time Resident of Moretown Passed Away

In the going of May S. Ward, widow of the late Hiram O. Ward, a woman of strong personality and great worker for all that was good and right in the home, community and church, has been taken. For a third time within a little over three years, neighbors, friends and relatives have gathered at the Ward home in Moretown in sorrowed respect at the passing of a member of the family. Three years ago this month, was the funeral of the father, a few weeks later that of the eldest son, Clinton, and last Sunday that of the mother.

May Ardelia Smith was born in Stockholm, N.Y., in Sept. 1848, the daughter of Harrison and Caroline (Canerdy) Smith. Losing her mother when quite young, as a young lady she came to live at the home of her uncle, Mark C. Canerdy in Duxbury, teaching in the rural schools of the town. It was while she was teaching at South Duxbury that she became acquainted with Hiram O. Ward, son of Deacon and Mrs. Earl Ward, whom she married. Living for many years in Duxbury she was closely identified with the work of the South Duxbury Church and later the Congregational Church at Waterbury.

While in the Phillip's District, their home was the community center and many who were then children remember the happy Christmas and New Years trees. As a member of the Ladies Union, at that time, she was a help. Because of increasing business interests in the Mad River Valley they moved to Moretown village and a few years ago built the beautiful home that so many have enjoyed with them. Here her hospitality and kindness have been without measure. All have been welcomed to the home and to any in sickness or sorrow she has been a strength. Her charities have been broad, but unassuming and in many cases, but only her minister and the one helped knew of the aid given. In the Methodist church in Moretown she has been a constant worker. As a Sunday school teacher, in the Missionary society and various church activities, her presence and influence has been active. In her home and her gatherings, her presence has been as a benediction.

Ill only a few days with pneumonia, nearly to the last her interest was keen and her faith in religious matters unbounded, her only concern being the salvation of others. Occasionally a character stands out pre-eminent. Such was that of Mrs. Ward who passed away last Thursday afternoon.

The deceased is survived by two sons, Burton S. Ward and Clair W. Ward of the Ward Lumber Company, Moretown; also four grandchildren over whose lives her influence will always rest. One sister and two brothers and two half sisters and one half brother survive. Mrs. C. E. Wynlan of Appenaug, R.I., E. C. Smith of Council Bluffs, Iowa; G. H. Smith of Evanston, Ill., Mrs. Luella Holly and Miss Demin Smith of Los Angeles, Cal., and Ezra Smith of Evanston Ill.

The house was filled Sunday afternoon. The arrangements were in charge of Thomas Ferris assisted by J.W. Bates. The Rev. Albert Abbott of the local church officiated, assisted by the Rev. Owen Gates of Cambridge, Mass. Beautiful tributes were given to the strength of her Christian character. E. G. Miller sang "Homeland" and "My Tasks," Mrs. Palmer accompanist. The burial was in the family lot in the South Duxbury Cemetery. The Rev. A. A. Mandigo of Johnson assisting the Rev. Albert Abbott. The bearers were nephews of Mrs. Ward's, Herbert Ward of Moretown, Jay Ward of Meredith, N.H., Carl Ward of Hanover, N.H., and the Rev. Owen Gates of Cambridge, Mass.

Among those present from out of town were Mrs. C.E. Wyman and Mrs. Frank Graves of Appenaug, R.I., the Rev. and Mrs. Owen Gates of Cambridge, Mass., Mr. and Mrs. Jay Ward or Meredith, N.H., Carl Ward of Hanover, N.H., Mr. and Mrs. William Ward and Miss Ethel Ward of Burlington, Mrs. Harland Munson, Levi Munson and Mrs. A. Munson of Morrisville; Mrs. Belle Frink of Yakima, Washington; Mr. and Mrs. J. C. Griggs of Barre and Mr. and Mrs. A. C. Huntley of Bolton. Large numbers were present from South Duxbury and Waitsfield as well as the home town.

Those who attended from here were Mrs. Edward Cullen, Mrs. B. J. Avery, Mr. and Mrs. B. F. Hart, Samuel Baird, Mr. and Mrs. B. R. Demerit, Mr. and Mrs. E. G. Miller, Mr. and Mrs. E. F. Palmer and Miss Annie Dorothy Palmer.

The Turbulent Years
1925–1940

In the 1920s the Ward family was beginning a new era. Burton was at the helm of the lumber company. He and his wife Annie Laurie Evans Ward lived in the Big House in the center of Moretown. Looking out across the street from their front porch—up to the left was the Ward Upper Mill, down to the right was the lumber

"First car owned [by Addie Sawyer] in Moretown. A couple of ladies in front of the "Big House." Oxen had to pull it in as it had a problem—rear end? 1915–1920 " —Burton Ward

company office in the Ward General Store, and even farther, out of view, stood the Ward Lower Mill. Burton and Annie's children were adults. Merlin and Kenneth, their sons, were living in Moretown, each taking on responsibilities within the company. Marion, their daughter, in the mid-1920s, was studying at Syracuse University.

For Vermonters, the national picture in the early 1920s was positive. Green Mountain native son Calvin Coolidge was in the White House. His taciturn manner and policies agreed with Vermonters, especially Republicans, and almost all Vermonters, including the Wards, were Republicans. "The business of America is business," Coolidge said in 1925, commenting on the nation's booming economy. Confidence in American prosperity was strong. In the 1920s real estate boom, hundreds of thousands of houses were built around the country. American families were buying home furnishings—dining room sets, pianos, Victrola record players in handsome wooden cases. Construction companies and furniture factories had great demand for lumber—and Ward Lumber Company was among the businesses supplying it.

New transportation and communication were changing American lifestyles, even in rural Vermont. Automobiles were no longer rarities—Model-T Fords cost about $300 and were within reach for middle-income families. Trucks were gaining popularity, too. Moretown, in the 1925 *Vermont Business Directory*, listed three automobile-related businesses. Where there were cars and trucks, people needed gasoline and repairmen. Radios were also rapidly becoming household fixtures, and radio stations were on the air in Vermont. (WDEV in Waterbury, founded in 1931, belonged to this the early wave of stations.) By 1925, according to the *Vermont Business Directory*, Merlin Ward was selling radios at the Ward General Store.

General stores of the era, the Ward General Store included, sold food, clothing, hardware, some furnishings, farm supplies, grain—just about everything needed in a small rural town. The Ward General Store, however, did not sell

*Burton's family: back row: Richard, Lois, Owen, Holly and Kenneth
Front row: Aline and "Granpa" Burton*

beer, wine, or other spirits. In 1919, Congress had ratified the 18th amendment to the United States Constitution, prohibiting the, "manufacture, sale, or transportation of intoxicating liquors within, the importation thereof into, or the exportation thereof from the United States...." Prohibition was in effect. Had Hiram Ward lived to see it, he would have been pleased; he had served on the local temperance movement committee. Burton was also a supporter of temperance. For decades after Prohibition ended, Moretown was still dry—people had to drive to Waitsfield or elsewhere to purchase alcoholic beverages.

The third generation Wards were coming into their own in the 1920s. Merlin, the eldest, born in 1897, graduated from the Mount Hermon School in Massachusetts in 1915. He studied at Syracuse University where he met fellow student Aline Hollopeter of Camden, New Jersey. They married in 1921. Richard, their first son, was born in 1924; two more children, Lois and Holly, would follow. Merlin managed the Ward General Store and from 1924, also served as Moretown Postmaster. For many years, Merlin and Aline lived in the Big House with Burton and Annie.

Kenneth Ward, born in 1900, continued his education at Dartmouth College after graduating from Montpelier Seminary. A member of Phi Gamma Delta, he played baseball and football at college and also competed on the wrestling team. Graduating from Dartmouth in 1924, Kenneth returned to Moretown and worked with Burton in management of the lumber company. Kenneth and Florence Miles of Sheffield, Vermont, married in 1924. Owen, their older son, was born in 1925; Wyman, their younger, was born in 1934.

The youngest of Burton and Annie's three children, Marion Lena Ward, was born in 1904. Like her brother Kenneth, after her elementary school years in Moretown, she studied at Montpelier Seminary. After graduating from Syracuse University in 1927, Marion taught and supervised the art program in the Walton, New York, schools. On August 21, 1931, she and Roland William Tweedie of Walton, New York, were married at the Moretown Methodist Church.

There was a lot to like about the early and mid 1920s—but as the decade started to wind down, ominous dark clouds of various kinds rolled in.

FLOOD OF 1927

On Thursday, November 3, 1927, two storms converged over Vermont—one from the south, one from the west. When these two rainmakers met, they stalled over the Green Mountains. For 38 hours the heavens opened. Northfield's weather station recorded a whopping 8.63 inches of precipitation—Moretown's soaking was probably similar. Some mountaintops are believed to have had as much as 15 inches in that short time. This extraordinary downpour caused the infamous Vermont Flood of 1927.

That October had been unusually wet and the ground was saturated. Rivers and streams were already running high. When the storms came, sheets of water flowed over open land, benign tributaries became raging torrents, and Vermont rivers went on a rampage. Water surged down streets, through houses, barns, and factories. Its enormous force carried away people, animals, buildings, trees, train tracks, electric lines, bridges, and fertile land. According to the Flood Survey Committee Report prepared for Governor John Weeks, 1258 highway bridges were destroyed or severely damaged. The flood took a terrible toll in Moretown.

The Ward family has documents that provide personal and compelling accounts of the flood. Burton and Kenneth were coming back from buying mill machinery in Morrisville, Vermont, when the flood started. A few days later, Burton wrote a detailed letter to his brother Clair who had made his home in California. Burton described their journey and the devastation in Moretown and beyond. As Burton and Kenneth were wending their way home, Merlin was tending the mills and store while Aline was at home. Aline and Merlin's second child was due that early November. She was at the Big House with her mother-in-law, Annie Evans Ward, when the rains started. Aline later wrote about her experiences on that day and night.

Burton Ward's letter to his brother Clair Ward:

> Nov. 15, 1927
>
> Dear Clair: —
>
> Your letter of Nov 9th just received.

Destroyed bridge going to Waterbury — Courtesty UVM *Special Collections*

A destroyed farm — Courtesy of the Moretown Historical Society

> On Nov. 2nd, at the end (10 p.m.) of a balmy day, it began to rain. The next morning Kenneth and I started for Lowell, Vt., up beyond Morrisville, as there was an auction of mill machinery and we needed a machine for stripping up dimension hardwood, like what we need to get out for Porter-Screen, and also squares.
>
> The rivers were just normal when we left home, but it rained all day but not hard enough to alarm me until we got back to Waterbury Ctr. And found the road giving away in the hollow below. We then started to go down trestle side from the Center and was (sic) told that the bridge that way had gone. We then drove down Little River and was (sic) stopped as the bridge had gone. We then turned to the Perry Hill road, thinking

that we could go around on hills and reach Waterbury, but after going through quiet stretches of water found another bridge gone; so we left our car and got a farmer to ford a roaring brook, now river, one at a time.

We made our way from there to Waterbury (4 miles) by road and fields and when we went by the old bridge below Mill Village one end was in the river. At Waterbury they told us that we could go no further as there was water in the streets so we could not get to either bridge. As the phones and electric lights were out we bought a flash light, which by the way only occasionally flashed, and started home by railroad to Middlesex.

When going by Edwards' Mill they were trying to get a family out of a second story window but could not and I think seven were swept away. There were several slides on the railroad between Waterbury and Middlesex and long cracks in the earth at Slip Hill several hundred feet of which went into the river soon after we passed.

On reaching Middlesex we found a twenty-foot deep washout in the south end of the railroad yard. We could not get anyone to carry us to Moretown as the roads were impassable so started on foot. The river was almost level with the road at the iron bridge and the waters made a most deafening sound as they poured over the dam.

When we got to Walter Turners (Knapp place, Walter bought it after his house burned) we asked him to take us as far as he could and on getting to the culvert Middlesex side of Edgerly house we found a deep wash out and car down under water with the driver inside, two other occupants were rescued. At Casey bridge [covered bridge just before the intersection of Rt. 100 and Rt. 2 in Middlesex just above the hydro dam] the water had cut deep channels both sides of the road and even with that extra channel was level with the bridge floor.

All the way, over the Common, we found slides and washouts where we had to cross a gully on guard rails. When we got to the village we found most of it vacated; many being at Will Griffin's (first house up the hill). Soon after arriving we heard loud cracking of timber, it was so dark you couldn't see a thing, and thought the bridge across the river had gone, the big clapboard shed, back of the Town Hall, with about two hundred thousand feet of good quality clapboards was, having gone before but in a few minutes we heard another cracking and this time it was the bridge. The other cracking was some of the buildings that floated away.

The water was so deep in the streets that we could not get home so asked some of the boys to get an axe and go up the brook towards Frank Battle's and fall a tree across the brook so we could get in back of our homes. Sleeper and South Hill bridges had gone, and after some experiences we came up into the top of Herbert Ward's and found him and Josephine also Mrs. Hazlett and Florence using hay for a bed.

We waded across the school house yard in the water to our knees, the water had gone down two feet then, and arrived home at just 3 AM.

Our house was full of refugees as most every one in our end of the village had left their homes. The waves rolled up on our lawn, within three feet of the house like ocean waves and down the entire length of the street except a short distance on hill at lower mill.

Property damage in Moretown Village.

At Don Fiedler's farm (H.J. Nelson) the water came into the second floor drowning fifteen head and a big pair of horses.

Bridge at upper mill taken. Mill house and barn, all wood, sheds, road, and 50,000 feet lumber in the upper mill yard, 24 ft. and Box-shop, just repaired this fall, all machinery and shafting, under box-shop torn out and into board-mill and river, broke in floor of saw mill and washed away all filling at end of mill leaving great hole.

Courtesy of the Moretown Historical Society

Ward Lumber Company's lumber and other debris stretched out for miles along the Winooski River all the way to Lake Champlain.

Moved Ernest Kingsbury's barber shop and road stand, tore off barn and back veranda of the Freeman house, leaving it in the road, took off end of creamery leaving it in our lower road, took creamery wood shed and wood, Freeman blacksmith-shop, old post office and garage, with three autos, (one Miss Whitney's, a good Star Coupster) others were old, also in the office building were quite a lot of goods, washed great hole in the back of sawyer place, took Evan's hen house and ice house also our back store, oat shed and salt shed, Bates' icehouse, horse barn and feed room at Griffith store, Mrs. Pierce's house and barn, barn and shed of the tenement where Paupino lived. Tea-house (Heath Place) wrecked, would have gone but was chained to trees. Undermined Frank Johnson's house (Arthur Neill) and moved barn against the Town Hall. Clapboard shed, bridge, across river, also lower mill dam and nearly a million feet of lumber in the lower yard with two-thirds of the village's supply of wood. Many of the cement piers for lumber were washed out. Moved Geo Parker's woodshed up into mill yard, took shed with clapboards from the old Russell Sawyer place (Clinton's) and ended up the village by taking the Ashley bridge.

The following houses, in the village, were under on the first floors and some on the second: E. Kingsbury (our old home), M.L. Freeman, Evans, H.G. Ward, Parsonage, Bates, House where Lucius lived when here, Fletcher Store, Jennie Pierce taken, Hayletts, Booth, Pappino tenement, Tea House, Sid Atkins, B.F. Griffith, Wilcox tenement up brook, Wilcox tenement next his store, Wilcox store, Powers house, hotel, Will Kingsbury, Chas Ashley, Frank Johnson, Will Hathaway, Geo Parker, and Tom Ferris place.

The Rutland Herald's *Extra edition for Sunday, November 6, 1927*

We lost about $5000.00 in store as we had nearly three cars of grain taken or soaked and all goods in back store, cellar, and on lower shelves in the store damaged.

Almost all our tenements are damaged to large extent as floors have heaved, paper and plaster ruined, and several chimneys gone. Almost all the furniture that got wet including pianos are practically ruined as they are coming to pieces.

The town lost thirty-five bridges, included in this number are four across Mad River and two across the Winooski River.

All lower Middlesex village was swept away or ruined. Miles Store, Creamery, and several other houses gone and a terrible gulf washed in the road; beginning at the old Nichols Store cellar around to iron bridge so there is another ravine nearly as deep as the river and wider before you get to the bridge.

At Duxbury Corner Mrs. Donovan (Town Clerk) house was washed away together with safe and books, and landed in Waterbury at end of where Winooski bridge was, many other houses from Duxbury Town Hall to Waterbury were taken, including the Arkley buildings in Moretown.

The main span of the new iron and cement bridge to Waterbury was carried down the river about one hundred feet and tipped bottom side up.

All of Waterbury was under water except from Smith & Somerville's store to the high school. (About ⅚ of the village) The water came up to pole at corner of old Wyman (?) store at top of hill and almost went over and was from the Duxbury Town Hall to sand hills back of Waterbury Depot, nearly ¾ of a mile.

Words utterly fail to give you any idea of the sights.

In Waterbury.

I saw all the pure-bred stock of Asylum, 117 head, as well as horses, hogs, etc which floated down river, in a grave waiting for lime to burn them up. The exact death list I am unable to give you.

Montpelier is in about the same condition. The day I was there you could not get a thing to eat or drink without going to the Red Cross rooms on Seminary Hill. Think of autos rolling down State & Main streets like pebbles in a brook. In some places there it was stated that the water was fifteen feet deep. The stock in all the stores was practically ruined.

The Central Vermont lost most of her bridges and much of her road-bed is gone so that the rails and ties for many miles are from one to fifteen feet in the air and in places even more from the ground. We do not expect to see a train for many weeks.

The record of this Vermont flood would fill a book as town after town suffered terribly.

The damage to meadows is immense from washouts and wash-ons. It is estimated that there are 10,000 loads of sand, dirt, stones, and gravel on Geo. Jones' nice meadow while the water made deep cuts in Feeley's, D. Bisbee's, Wallis', D. Fletcher's, and many others.

Our loss will be from $50,000 to $75,000 according to cost of new dam. We are salvaging some of the lumber but at such expense that we will not realize much.

There is a dam of lumber, wood, bridges, houses, tree, sand and muck nearly eight feet high, one hundred feet wide, and eight hundred feet long across the lower end of Murray's meadow.

Will Conrad's meadow damage from wash will be about $500.00. Will Shepard lost eight head of young stock and hay so wet that it heated to such an extent that he feared it would burn as Len Jones' did at Middlesex, and he watched it day and night for some time. I expect that his hay is ruined. I told him that I would help him out on hay if it was. I have been trembling with fear that he would throw farm back on us.

The Northfield weather bureau reported 8½ inches rain in thirty hours; according to that, if Dr. Bates and I figured correctly, 12,000,000,000 pounds of water fell on the town of Moretown alone.

Both mills stayed on foundations as they were on high cement piers. We have put about a carload of cement in walls and piers at upper mill this last year; beginning with a solid cement wall on one side and a very high cement pier in center and part cement wall in front.

Another great loss, hard to retrieve, is the morale of the people. Many of the flood victims owned nothing save their furniture and perhaps a piano and to see that all fall to pieces is certainly trying.

My courage is not the best for as I told Annie, not more than two months ago, we had bought water-wheels (3), built and repaired mills, improved tenements until we had things in very fair shape and now everything is all shot.

It will take six weeks to get the upper mill in running order and it will be ten months before we could get the dam in below if we should decide to build.

Have you any suggestions to make in regard to Will Shepard?

Very sincerely yours,

Burton
BSM/MW.

The inscription on the back of this photo reads:

"The Moretown General Store. It was badly damaged by the Flood of 1927.

"This building was bought by Hiram Ward about 1889 who was operating 2 lumber mills in Moretown then. At that time the ground floor was a Ward Lumber Co. store. In 1893 the upper floor was made into an apartment, later to become the Lumber Office. It was burned to the ground in 1965."

Aline Ward's account, titled, "The Flood, Vermont, 1927"

They sat looking out of the big picture window on the second floor of the large rambling house; the young mother, Aline, large with the baby expected daily, holding the three-year-old, Richard, on her lap. Sitting beside the two, was Annie Ward, mother-in-law of Aline.

This was November 3, 1927, in the little Village of Moretown, Vermont. The "Village" was a small section of the Town of Moretown that extended along the scenic Route of "Route 100" twelve miles from Montpelier, the Capital of the State, for about one mile. It formed a semi-circle and was spanned by a bridge over the "Mad River" at the North and South ends. There was a Lumber Mill at each bridge, for the water of the Mad River furnished the "Power" to run the Mills via of separate dams. So the Village of Moretown combined it's lively-hood and maintenance within this small section. The Hills rose in back of the Village to the East, and there were farms and homes there, but the "front" of the Town was the "River." The Elementary

Courtesty UVM Special Collections

Waterbury just after the flood. The note reads: Flood ruins, Waterbury, Vt, Nov 5th 1927, Huard Photo

Courtesty UVM Special Collections

Watching from Stowe Street in Waterbury as the fire house drifts along Main Street in the background

School, two Churches, two stores and Town Clerk's office were in this narrow way.

The House where the little group sat in the Window, was the home of the Ward family who operated the Mills of the Town. Of course, the men of the family were busy at work.

So what was "special" about "looking out of the window?" Well, it was raining; had been raining all morning, and the night before. But it had rained before—no call to be unduly alarmed. Of course, little Richard could not go outside to play The Mad River was normally a quiet lazy stream, flowing placidly thru the Green Mountains to join the large Winooski River at Middlesex, six miles away.

So—time dragged along on this November 3rd. It was now 2 o'clock in the afternoon. It continued to rain. Annie said; "I do believe the River is over the Bank, and beginning to crawl over the field. No worry" this was a vacant lot, a flood plain about a quarter of a mile wide, leading to a steep bank to Route 100.

There were houses on the river side of the highway, but the Big House was on the East side and on slightly higher ground.

Jokingly, Aline said: "I hope I don't have to go to the Hospital to-night to have the baby. Anyway, I have my bag packed." There was no Doctor in the Town of Moretown, so one had to travel 12 miles to Montpelier, using one bridge at the North end of the Village and another, much wider, at Middlesex, six miles away. But surely, this was a remote possibility.

But something was beginning to happen. The men of the little family, Burton the Senior and Merlin the young father, checked in the house—everything O.K.?

Burton was going from Mill to Mill, was watching the Mill machinery, so close to the River; the dams were a raging torrent. Merlin was at the family owned General Store, getting some of the stock of heavy boots and shoes moved up out of the cellar. The Store was on the River side of the Village and water was pretty close to seeping in.

It was 3.30 p.m. The Mill whistles blew; whose who lived up the Hill, hurried home, as did those of the Village to try to barricade their cellars. And still it rained! School was dismissed at 4 p.m.—but there was a problem!! The ends of the Village were lower than the center, and the water was over the road, a foot and more deep. The children could not get through!

The Big House was on the highest level of the ground, so it was the only place for the children to go. 37 children were herded into the house, some of them noisy with the thrill of adventure, but more of them frightened that they could not get home. The phone began a continuous ringing as anxious parents wanted to know their children. But in half an hour, the phone went "dead." The line was down. Little Richard was delighted to have so many play-mates; Annie and Aline were concerned about how they could feed so many and there loomed the possibility they would need to stay over-night!! Burton and Merlin checked in again, but struggled out to try to chain down machinery. The River was raging thru the Mill Yards.

Five o'clock and still it rained!! The entire families from the homes in the lower sections of the Village came floundering into the Big House, bringing their tiny children, and in some cases invalid older family members.

At 6 p.m. Merlin and Burton, exhausted and defeated, came in to say "The upper Dam has collapsed as well as the Bridge," and they could no longer reach the Lower Mill, the water was 6 feet deep there.

And it continued to rain!! The men who were marooned in the House, gathered on the porch at the front of the house, some 25 or more of them. Every 15 minutes they measured the water off the step of the front porch. If it reached 12 inches they would have to evacuate the house by climbing the Hill in back of the house, a Herculean task for those with tiny children and older folks; not to mention Aline with her expected baby! A few of the families had brought a bit of food and milk, but every child had been given something to eat. They were now sitting around the two large living rooms, on chairs or the floor. A few of the little children were asleep.

At 9 p.m. a shout went up from the porch. It had stopped raining!! Another half hour and a pale moon was trying to break thru. The water about the House was waist deep, but the house was built on a high foundation. There was a foot of water in the cellar. The Mad River deluge was still rushing down thru the Village, and continued unabated until 2 a.m., as it gathered force thru the narrow mountain

Courtesty UVM Special Collections

"*A wrecked home in flood stricken Waterbury, VT, Nov 4th, 1927*"

Courtesty UVM Special Collections

"*Flood wrecked Central Vermont R.R. Bridge. Montpelier Junction Montpelier, VT. Nov. 4th, 1927*"

Valley. But it did not rise and by 2.30 there was a slight drop in the level. By 4 a.m. a few hardy fathers struggled down the hillside in back of the house to gather their children.

At day-break, those of the Village who could, sloshed their way thru the water and debris to go to their homes to see destruction and devastation; in some cases, it reached to the second floor.

Both bridges were gone, the Mill yards were a shamble of battered lumber and logs that had been orderly stacks of Lumber to be shipped and logs to be sawed. The bridge at Middlesex had been completely washed away. The river at that point was twice as

Courtesy UVM Special Collections

The Bolton railroad station standing behind the wrecked tracks

Courtesy UVM Special Collections

"Flood wrecked Central Vermont R.R. below Silver Ledge Hill, Montpelier, VT November 4, 1927"

wide, forming two channels around the rocks. Moretown was completely isolated from Route 100. The only access and contact with a source of necessary supplies or services was Northfield, reached by traveling a muddy dirt road "over the mountain," 10 miles away. And people did immediately come to Moretown from Northfield, some of the students at Norwich University, came by horse-back, with food and supplies.

And now, Aline and her family were indeed worried that the "Baby" would come. It was a week before access across the two open passes was established. Swaying foot bridges connected Moretown Village with Route 100, but to get to Montpelier and the Hospital, there was still that widened span at Middlesex. But at the end of the week, November 10, Merlin took Aline by car, over the Hill in back of the Town on the Moretown Common Road. This was normally a fair gravel road, but now a mass of mud that scraped the bottom of the car. At Middlesex, they left the car, struggled down a ladder to the river bed, up a ladder to the top of a rock in the middle, down and up two more ladders to reach Middlesex. Here Merlin prevailed upon a man to let him have a car to go on to Montpelier.

As they reached Montpelier, it too was a scene of devastation and had been placed under military guard. They let the car thru to Heaton Hospital, situated on a Hill just outside the City. There Merlin managed to get a room for Aline, though it was crowded with evacuees of the City. Aline's Doctor's own home in the City had been ravaged by the Flood. The phones were still "out." Merlin left to struggle back to Moretown to the gigantic problems of salvaging a business and homes.

After two days of waiting in the Hospital, a room was found in the City for Aline; they needed the room in the Hospital. There she waited for 10 days until November 30, when the need to go to the Hospital became urgent. She called a Taxi, and before long Lois Evans Ward was born. Merlin was called on the then functioning phone and informed he had a daughter. Survival was complete!!

The Great Flood of Vermont on November 3, 1927 has been a land-mark date for this family.

It has been a Day of Memories for many Vermonters, one of destruction and loss, but one that brought out great courage and fortitude to endure. It also showed great compassion on the part of many who escaped much of the destruction.

Beyond the Ward family, an account written by Luther B. Johnson, "The Floodtide of 1927," offers another look at the 1927 devastation.

"Moretown, further down, suffered heavily, losing nearly 20 bridges and much highway. In Moretown village three houses with barns and shop were lost. Ward's upper mill was practically destroyed and the dam washed out at his lower mill and about a million feet of lumber, including a large amount of dressed clapboards, floated away. The lower part of nearly every house in the village was filled with mud and water. Both of the bridges across the Mad River above and below the village went out. One side of the dam at the power plant on the road to Middlesex succumbed. The Lane plant and dam and the Middlesex dam and plant all went." [p. 104.]

Etta Johnson of Moretown wrote a poem about the deluge and the toll it took on the community.

> In the early days of November
> Nineteen hundred twenty-seven
> A most unusual downpour
> Was sent from out of heaven….
>
> My tale is one of the many
> But of one little village I love
> It lies deep in the valley
> My home on the hill above…

She tells how the farmers suffered through loss of land and animals, and the difficulties of not being able to transport their milk and cream out or grain in. She writes of the families who lost their village homes including that "Miss Whitney was wise and had hers chained, So that house didn't go."

> Ward Lumber Co. the heaviest loser
> Property damage, grain, and lumber and wood
> With Moretown depending on them for work
> We pray that their courage keeps good.

As soon as the floodwaters receded, recovery and reconstruction began. Calvin Coolidge visited his home state in 1928—the devastation of the flood still evident. He famously concluded one speech with the following.

Courtesty UVM Special Collections

"Flood wrecked Central Vermont R.R. Bridge below Depot Montpelier, VT. Nov. 4th, 1927"

Courtesty UVM Special Collections

Flood wreckage, residential section, Montpelier, VT. Nov. 5th, 1927

Courtesty UVM Special Collections

Courtesty UVM Special Collections

"I love Vermont because of her hills and valleys, her scenery and invigorating climate, but most of all because of her indomitable people. They are a race of pioneers who have almost beggared themselves to serve others. If the spirit of liberty should vanish in other parts of the Union, and support of our institutions should languish, it could all be replenished from the generous store held by the people of this brave little state of Vermont."

Massive bridge and highway reconstruction started right after the flood. Vermont sold bonds to help finance the needed infrastructure investments. The federal government also provided flood relief funds. The construction seasons of the next two years were bustling as new concrete and steel girder bridges replaced destroyed wooden and covered ones. Roads were rebuilt with an eye to needs of automobile and truck traffic.

Even with their extensive loss of lumber and damage to dams and mills, the Wards repaired their facilities and resumed production. Many mill owners never recovered from the flood. Upstream in Warren, the floodwater closed several mills forever.

Depression In Vermont

Courtesty UVM Special Collections

Reconstruction and flood recovery were still underway when a new disaster struck—this time on a national and international scale. The boom times of the early 1920s ended. Many factors, including the Wall Street Crash of October 1929, combined to cause the Great Depression. New home construction around the nation all but stopped; markets for home furnishings collapsed. Unemployment soared. Ward Lumber Company was hit hard by the economic downturn. Firms that had been buying hardwood for furniture or softwood for construction went out of business, surviving companies placed much smaller orders.

Residents of rural Vermont, including the Wards, had considerable resilience.

When the Vermont Folklife Center conducted interviews in 1992 of dozens of Mad River watershed residents, many older community members still had razor-sharp memories of the Great Depression. Several interviewees recalled the Depression as a time of hardship, when there was little money. Overall, though, there was a sense that people got by. Several Moretown residents who were interviewed recalled Ward Lumber Company's importance to the community during those years. At times, the hours of work were reduced, but the company kept its employees. Along with the mills, work continued in the plantations. Loggers still cut trees and teamsters still hauled them to the mills, even if not as many or as often as before. The Ward General Store continued as a source of supplies of daily life. For some families, the store also provided some income, purchasing eggs and other goods.

As the editor of *The Vermonter* magazine noted about Burton Ward in 1931:

> During the recent depression the mills have been kept running by rigid economy and close attention to costs, thus employment has been furnished to practically all the workers of the village and for several from adjoining towns. Living costs are very reasonable as Mr. Ward furnishes many with homes and fuel at about one-half the price asked in larger communities.
>
> Good fortune seems to have attended the people of Moretown.

The Depression stretched out for over a decade. The Wards who were adults in that era are long gone, but even decades later, Owen Ward recalls hearing that Ward Lumber was borrowing money to get by.

Fire

In the early hours of a September, 1935, morning, flames lit up the Moretown sky. Fire was always the enemy of lumber yards—once a blaze started, the stacks of drying boards provided ready fuel for the conflagration.

Even in 2011, Owen Ward vividly recalls the flames and the fear during that late night inferno. Kenneth and Florence Ward and their two children lived in the house next door to the Moretown Elementary School. A large window in its kitchen faced north. Standing with his mother in the kitchen, Owen could see the flames and smell the smoke. He was only ten years old at the time, and decades later, he still recalled the genuine fear of not knowing if the town or the company would survive.

Fire Today Destroyed Ward Lumber Company Mill at Moretown.

Waterbury and Montpelier Fire Departments Called to Protect Town and Mill Yard—Unknown Origin—Less Than Ten Thousand Feet of Lumber Lost Out Of Two Million in Yard.

Fire created one of the greatest losses to be suffered in the Mad River Valley in recent years, early Thursday morning of this week, when the "lower mill" of the Ward Lumber Company in Moretown was completely destroyed.

Fire broke out from an undetermined source between 2:00 and 2:30 and was not discovered by residents living close by until the flames had made great headway through the mill.

Two Departments Called

Fire departments were summoned from Montpelier and Waterbury and with one at either end of the blaze which covered a wide area the flames were confined for the most part to the mill proper and the greater part of the mill yard which contained some two million feet of stacked lumber was not damaged. While the mill itself was a complete loss, estimates place the amount of damaged lumber at between 5,000 and 10,000 feet.

The Ward mill was one of the finest in this section of Vermont, much new machinery having been installed there, of which the latest was a new and expensive planing machine.

The Montpelier Fire Department sent a pumper and five men, who responded promptly and set up at the lower side of the blaze on the downstream side of the mill between the fire and a large part of the well-filled lumber yard. The streams were put on the blaze from this end doing very effective work.

Made Quick Run

Twenty-five minutes after the call in Waterbury and about two minutes after the Montpelier outfit moved in, the Waterbury pumper arrived and went into action with six men at the upper end of the mill property between the fire and the town. Two more lines were laid which were used to extinguish the fire at the mill itself, protecting the penstock from the dam, and to put out fire which had crept into the log piles nearby.

Both departments put in nearly four hours at the fire before picking up.

Third Great Loss

The fire this morning at the Ward Lumber Company is the third major disaster which has befallen the company since it has served Vermont builders from one of the greatest lumbering sections of the state. Another great fire loss was sustained in 1895, and in 1927 the Vermont flood wrecked both of the company's mills and carried hundreds of thousands of feet of lumber down the Mad River and left it strewn the entire length of the Winooski valley.

Moretown Pump Smashed

A small pump owned by the town of Moretown and carried on a truck, was smashed beyond use when it parted company with the truck at the bridge in the streets of Moretown, while being rushed to the fire. The town was left without an effective water supply for fire until the Montpelier and Waterbury Departments arrived.

Source *Waterbury-Stowe* newspaper September 11, 1935.

Soon, the cause of the fire was determined. As reported in the *Waterbury-Stowe* newspaper of Wednesday, November 13, 1935.

POTVIN ARRESTED FOR ARSON

**Waitsfield Man Pleads Innocent;
Charged With Burning Ward's Mill**

Rudolph Potvin, 30, farmer of Waitsfield, pleaded not guilty to arson in Montpelier city court Friday morning. Judge A. C. Theriault ordered Potvin committed to jail for want of bail, set at $2,000.

Potvin was arrested by Sheriff Henry C. Lawson Friday morning on a complaint issued by State's Attorney Webster E. Miller. The complaint alleges that Potvin, on the night of September 12, maliciously and with intent to burn, set fire to the lumber mill of Burton S. Ward of Moretown.

Potvin's arrest was largely due to the long and careful investigation made by Deputy Fire Marshall Frank Eaton of Waterbury, and by Sheriff Lawson. Mr. Eaton commenced his work immediately after the fire which caused a property loss estimated at $12,000 and destroyed a mill which employed forty men. The two officers diligently followed every lead until they had covered the entire ground.

Potvin, when arraigned in court, was a bit temperamental, refusing to get a lawyer, maintaining his innocence and threatening to 'put a piece in the paper' after his innocence had been established. Judge Theriault explained to him his rights under the law, telling him he was entitled to an early hearing in the municipal court to determine whether or not the evidence warranted his being bound over to be tried by the county court.

Judge Theriault stated at the close of the court that the binding over hearing would probably be held early this week.

Once again, the Wards rebuilt the mill.

Hurricane of 1938

If flood, fire, and Depression all in the space of eight years were not enough, in late September, the Hurricane of 1938 rolled in. The first hurricane to hit the Northeast straight on since 1869, this monster storm hammered New England and laid another blow on Vermont. It damaged or destroyed more than 57,000 homes, killed over 650 people, and devastated transportation systems. Along with pounding rains and massive flooding, this storm also had powerful winds. Every New England state was hit hard—serious hurricane damage extended from Maryland to Quebec.

In Vermont, the hurricane killed five people. In some ways it was a replay of 1927. Houses, barns, farmland, and over 2,000 miles of roads were damaged. Power lines were downed. A radio transmitting tower of WDEV in Waterbury was knocked down and out. Floodwaters again charged through communities including Moretown. Lumber from the Ward yards was carried downstream as far as Lake Champlain.

A Ward family scrap book includes photos and memories of this second deluge.

Flood and Hurricane of Sept. 21, 1938 from Family Scrap Book:

> "Cited by the press to be the most damaging catastrophe that ever occurred on this continent. Billions of feet of lumber was fallen, millions of which will not be salvaged although the gov't has hundreds of mills in operation. Many of our New England ponds and lakes are filled with white pine logs which need the water for preservation. Damage limited to New England states where on sea shore hundreds of resorts were washed to sea along with lives lost and many buildings ruined inland. Hundreds of farms ruined through loss of sugar places and timber holdings. Following are a few pictures showing local damage in which the Ward Lbr Co lost many thousands of dollars."

Damage at the Merrill Reagan house

Lower mill damage

Picture taken below the dam at upper mill the day after the flood. Note washout at the end of the mill and the wagon nearly gone.

From the Ward Family Album

"Stacks of lumber at the lower mill. The washout was many feet wide. About 40 sheds of dry hardwood are lost and others are flattened as shown."

Devastation at the lower mill, inscription from the Ward album page:

…showing local damage in which the Ward Lumber Co. lost many thousands of dollars. Showing lumber, wood, and rubbish from our lower mill washed onto land owned by Mrs. Will Hathaway. 200,000 ft of lumber was entirely lost down the river. 700,000 ft of lumber was badly thrown around …

Brighter Side of the 1930s

The years from 1925 to 1940 certainly dealt abundant challenges to Moretown. In spite of considerable losses, Ward Lumber Company survived the disasters and Depression. As the company pulled through, many in the community depended on Ward paychecks or income from selling timber to the mills. For some families, a little extra income from selling eggs to the Ward General Store was welcome help. Yet, even with the hardships, those years were not all drudgery. A strong sense of community and the resources and joys of rural Vermont provided good experiences even in those tough times.

During the 1930s, the fourth generation of Wards—Owen, Wyman, Richard, Lois, and Holly—were children. Their grandfather and fathers dealt with business—the plantations, mills, and store. The young Wards studied at the Moretown Elementary School and kept up regular attendance at church. (Burton, like his father Hiram, was a pillar of the Methodist Church. A tap of his hand on a knee quickly silenced any fidgeting grandson.)

Summer and winter, the Mad River offered abundant diversions. A terrific swimming hole, complete with a big rock for jumping, was just a short walk from the village. Fishing holes often yielded a good catch. In the winter, occasionally the ice was good for skating on the mill pond. Away from the river, the village baseball diamond was a center of summer recreation. In winter, the hills provided great terrain for sledding and even skiing, although there was no tow.

The Grandchildren: Richard, Lois, Holly, Wyman, and Owen

For the Ward grandchildren, as for other young people in Moretown, life was not all swimming holes and baseball. This fourth generation, like the generations before them, were raised with a strong work ethic. The young Wards' responsibilities included helping with Burton's Guernsey cattle, haying, and pruning in the plantations.

Burton's Guernseys

"Here Boss. Here Boss." Burton Ward called to his fawn and white Guernsey cows, and the Guernseys would often listen, especially if it was milking time.

For many years, Burton Ward kept a herd of Guernseys. He was a great fan of this breed known for producing high-protein, high-butterfat milk. "Golden Guernsey" milk was sold as a premium product. Most of Burton's cattle were at the Flannagan Farm up by Moretown Common. In addition, he generally kept a couple of cows by the Big House in the village.

In the summer, Burton pastured some of the herd in Fayston, up the road later known for the Fly-In air strip. After the heifers had been there for several weeks, they needed to be returned to the farm by the Common. For the transfer, the cattle were herded several miles—from the farm, down Route 100, through Moretown village, and up to the Common. Burton's grandsons were often enlisted for the drive—a very long day.

Merlin in back and Kenneth Ward—at the farm with cows in 1907

Burton and one of his Guernsey cows on the farm

Burton Ward's farm on Paddy Hill in Moretown, where the cows where kept. It was farmed by Henry Nelson, his wife and their nine children. Sometime in the 1940s, a fire started at the Goss farm nearby. The newspaper reported that "the wind carried embers to the Ward/Nelson farm. Both prosperous farms burned to the ground, killing Burton's yearlings and young calves, a loss Burton described as, 'a calamity to a small town…'"

BASEBALL!

Baseball, the Great American Pastime, was Moretown's sport. Along with the rest of the nation, Moretown residents enthusiastically played baseball, attended games, and cheered on their teams. Baseball's popularity in Vermont dates back to the late 1800s—the early Northern League had semi-professional teams in many Green Mountain communities before the 20th century. The University of Vermont team was renowned, they even competed in a collegiate national series in Chicago in 1893.

The community likely had earlier teams, a photograph from the Moretown Historical Society of the circa 1915 team confirms the sport's local following by that date. Among the handsome young men, only George Evans is identified. A few words on the back note that he was called away from the pitching mound mid-game on July 5, 1915 for the birth of his daughter Lucille. (George Evans and Annie Laurie Evans Ward were siblings.)

Although the exact dates are unclear, for a time Burton Ward supported a semi-pro team in Moretown, a team that included collegiate players. The "pro" part was far more modest in those days before multi-million dollar contracts, but local excitement ran high. The Moretown baseball field was behind the Main Street houses. Fans sat on the slope to watch the games. Most of the games were likely against Vermont teams—Montpelier and Barre had teams, so did Northfield, Burlington, of course, and others.

A huge crowd—somewhere between 1,500 and 5,000 according to local lore—turned out for at least one Moretown game. The local team played the Mohawk Giants, a team of Black players from New York. Accounts in the Moretown Historical Society archives note, "The Moretown team was active about 5 years and they were good!!" Team members for at least part of the time included: Cornelius (Coriolano) "Kio" Granai, Kenneth Ward, Milford Grandfield, and fellows with last names Wheaton, Shontell, and Atkins. The team reportedly played games on Saturdays but never on Sundays. Burton was very devout; the Sabbath was for church.

Courtesy of the Moretown Historical Society

Circa 1915 — Moretown's baseball team

Moretown's Kio Granai made it to the big leagues, and played for the Detroit Tigers, one of the top teams of the time. Granai's connection to Moretown extended beyond baseball. Born in 1897, one of eleven children of an immigrant family from Carrera, Italy, Kio moved to Vermont in 1902, apparently when his father came to work in the granite quarries. Kenneth and Marion Ward met Kio when they were studying at Montpelier Seminary, around 1914. From their friendship, Kio began working for Burton, logging and helping with dam repairs. Burton and Annie Ward became like second parents to him; he is listed as a "boarder" in their household on the 1920 census. The Wards supported Kio in continuing his education at Syracuse University and in Syracuse Law School. Kio Granai, more formally Cornelius Granai, became Vermont's first Italian American State's Attorney. He served as mayor of Barre for many years and represented Barre for six terms in the legislature. A distinguished Vermont statesman, Granai kept his ties to the Wards and Moretown through decades.

Kenneth Ward was also a talented ball player. Beyond local games, he went on to play on the Dartmouth College team. After graduating, when he returned to Moretown, he was again at home on the baseball diamond in the village. Kenneth also shared his skill and enthusiasm coaching a new generation of boys. One Dartmouth newsletter from 1937 included this tidbit about Kenneth, "The first of his two sons, Owen 12 and Wyman 3, unquestionably plays on his father's ball team the Moretown Sluggers, consisting of boys from eight to fourteen."

The big games and semi-pro baseball days were exhilarating, but baseball thrived on many levels. Games in Montpelier, Burlington, and beyond attracted Moretown fans. Boys and men played on a range of local teams. Owen Ward recalls countless summer hours spent playing and practicing—his skills honed in Moretown led to a place on his New Hampton School varsity team. Through many summers, teams in Mad River Valley towns kept up a schedule of competition.

Kenneth Ward and his Moretown Slugger teammate, Nelson Duba

In & Around Moretown

While the history of Ward Lumber Company is the story of four generations of one Vermont family, it is also the story of the hundreds of men—and women—who worked for and with them. In some families, fathers, sons, and grandsons all worked in the mills or the forests over the years. While lumbering was largely a man's world through those years, women also worked for and with the Wards, including bookkeeping in the office and tending the store.

"We had so many wonderful and faithful employees over the years," Owen Ward recalls. "We couldn't have been the business we were without them."

Courtesy of the Moretown Historical Society
The first Ward IGA grocery store, in 1930, eventually known as the Upper Store. The post office is the on the left.

The Ward Stores

General stores were social and supply centers of rural Vermont towns. With food, clothing, tools, grain, the newspaper, sewing supplies and much more, the store was the all-purpose source for local families' needs. From the beginning of the 20th century, the Wards were proprietors of one of Moretown's two village stores.

The original Ward store was located across the street from the Methodist Church. For many years, groceries and house wares were in the front, hardware and the grain room were in the back. Meeting its community's needs, the store sold radios when they became popular in the 1920s. Installing gas pumps out front, the Wards accommodated the motoring public, selling Gulf gasoline.

Merlin Ward, along with his involvement with the lumber business, managed the store. He also served as Moretown's Postmaster from 1924 until 1963. Conveniently, the post office shared the store's space; it was on the left side of the building. With this combination, just about everyone in town spent time there. They did not just shop and leave—this was a public place to meet and catch up with friends and neighbors.

Men who worked for Ward Lumber, and independent jobbers who worked with them often lingered—telling stories, pulling occasional practical jokes. Many Ward employees had accounts at the store and they could charge their purchases against their wages.

The Ward Lumber Company office was upstairs. The bookkeeping, management, sales, and purchasing all were based here.

In the 1950s, the Wards purchased Moretown's other store, located across from the Town Hall. The company moved the grocery business, post office, and company offices there, leaving the original building as the hardware store. The building that was home to the relocated Ward IGA Store continues today as the Moretown General Store.

Vermont Folklife Center interviews

In 1992, the Vermont Folklife Center based in Middlebury undertook an oral history project with the Friends of the Mad River, a local watershed organization. More than 50 interviews with long-time Valley residents were conducted, recorded and transcribed. From this project, an audio documentary, *Mad River Valley, Crucible of Change,* was developed. This project preserves residents' recollections of the economic, social, and natural history of the area. These excerpts from the interviews give us glimpses of Moretown and the Ward Lumber Company's role in the community.

They had everything here. They had grain, and clothing of all kinds, winter, summer. You could buy anything.... It was really wonderful.

At quitting time all the mill help used to come into the store and do their shopping. And it seemed as if they lived from day to day. Now they probably didn't. But they used to come in and they would buy their groceries for that meal I should say. Things were charged, they were put on the slips. You grew to know everybody. There were very large families. In fact, I think there were three families in town with ten children. But nice families, and they were, everybody was close.

This was one big family. Everybody knew everybody. It really was great. Nothing like it. My children grew up when it was the best of times I say... All there was in town was the mills. And every family worked at the mills.

— Frena Cutler, who worked at one of the stores

I remember the store, the Ward's store. They had just the upper store at that time, which was opposite the church and it burned...You go into the store and there's these old-fashion registers. And the guys used to play cards there. On Saturday night they'd all gather around, play cards.

You sort of had to wade yourself in between. Then they had all this stuff piled on. I guess some of the stores still do that, sometimes you see all the coats and everything piled right on the shelves? We'd go down, mother sent me down for groceries, too. Mother used to sell eggs to the grocery here in Moretown. With the money she got from the grocery, I mean for the eggs she got paid for most of our grocery bill, which was really a godsend, especially in the depression era.

— Lettie Conrad about the Ward Store

When I was growing up, we used to have the box socials at the Town Hall. The Wards, of course, having the businesses, used to have the community Christmas tree, a night or two before Christmas. The school children would put on a program. The Wards would get these two huge great big trees and have them set up in the town hall, and they'd be decorated. They'd put gloves and things on the trees for the men. The children, every child in town, got a candy box, if nothing else. That sticks in my mind as the way they did back then, back in, oh, '43 or '44.

— Wilma Maynard

They didn't pay big wages, but during the depression, when you couldn't find work anywhere else, they kept their men and people working where in many, many cases, people just did not have jobs. And, I know that there's a lot of things that the company did that, and Wards did that, they never got much credit for. They didn't blow their own horn, in other words.

— Mary Reagan

In the wintertime, people would put their car up and they would go to town with their horse and buggy. A lot of the older people that didn't know how to drive would use horse power to get where they were going, you know.

— Robert Wimble

Moretown was a better place, probably, to live in a way; the Ward Lumber Company was here doing business. They didn't always pay the highest wages but people in Moretown, most everybody had a job, other than farming. There were some farms here but the Ward Lumber Company had three mills and they did a real big business.

— Bob Gove

It has been remodeled, and the post office moved into its own building across the street. The Wards sold the store in 1967.

The original Ward Store continued as the hardware store for several years, then was used for storage by the company for almost five years while the top of the building was rented out as a large apartment to the Mann family, a couple with eight children. In April, 1964 a spectacular fire destroyed the building, started a forest fire and damaged several homes nearby requiring 150 firefighters from five towns as well as townspeople to put the fire out. No injuries or fatalities were recorded, but the loss to the Wards was considerable. The site was leveled and became a lawn for an adjacent home.

Courtesy of the Moretown Historical Society

"...at the hitching post" is inscribed on the back of the photo. The Moretown Elementary School is the low building in the back.

TO OUR GOOD CUSTOMERS:

After a lot of talk about it, we have finally decided to sell our store. Ward Lumber Company has operated this store in Moretown Village for about 70 years but time marches on and we must make a change.

On Monday morning your new storekeepers will be Russell and Dorothy Bombard from Barre who will move into the apartment with their children after school is out. Russell is very well able to run a nice store for Moretown. He has worked for the A & P for many years.

We thank you sincerely for your trade over the years and hope you will give the Bombards all the support you can. The lack of a good store in the village would have a very bad affect on the value of everyone's property here. People interested in buying property in our town would be very much concerned with the store situation.

May 18, 1967   WARD LUMBER COMPANY, INC.

MORETOWN METHODIST CHURCH

Hiram and May Ward helped establish, and were early supporters of the Duxbury Church, but, services weren't always held there. From H.O.'s diary in 1874, he notes that he often observed the Sabbath in more than just their little Duxbury church. There are many notations telling of frequent attendance at the Moretown church, and occasionally in Waitsfield.

After the move to Moretown, they and future generations of the Ward family became integral members of the Methodist Church.

The booklet, *A Brief History of Moretown, Vermont 1982*, highlights a contribution of the Ward Lumber Company to the Methodist Church, which had been built in 1853 and was being renovated in 1922:

In 1922 several of the faithful stewards of the church decided the building was showing the ravages of too many years of hard use. Not just church sessions, but Sunday School, Epworth League meetings, Christmas tree celebrations, and other group meetings were held there. That summer, meetings were held in the Town Hall so that renovations could be completed. All new pews were installed as well as new areas for the choir and chancel. The unusual feature of the remodeling was that the lumber for the work was taken from the birch trees on the Ward Lumber Company land in the town. The lumber was sent to Burlington for the carpentry and then installed.

A view of Moretown looking down at the former Dr. Haylett's home (now the Schultzs) with the two churches in the background. The Methodist Church is the one on the left.

Looking back on maple sugaring seasons

They had the maple sugar on snow party in the spring. They would go and get the ice, get the snow, sometimes they would go to Granville woods to get the snow. The farmers would donate the maple syrup. The ladies of the church would boil it down for sugar on snow and there would be plenty of pickled eggs and sour pickles and raised donuts. Haddy Bates was great with the raised donuts and Helen Austin of course. It seems like, Mr. B. S. Ward, that was the father of the Lumber Company, would underwrite the cost of most of the young people going to the supper if they didn't have enough money. That was nice for the youngsters to be able to do that. It was a real treat, you know, it was sort of a festival of spring. [There were] quite a few sugar places around and in the spring, and we had, of course, what we call mud season, where, which shut the school down for three weeks and while the people and the youngsters sugared. They couldn't go through the mud anyway. Well, I guess they could walk, and most people walked to school anyway, but there was a three-week vacation and it would coincide with sugaring, so we called it sugaring vacation.

— Robert Wimble (VFC excerpt)

The former lower Ward IGA Store, now the Moretown General Store and gas station

[At the town hall] they had a community Christmas tree. The Ward Lumber Co. always gave all the men gloves or something, and the trees were trimmed with that. And the women were given aprons or kitchen towels or some item for the home. And then the families brought gifts for their friends and their children… The school children put on a program. They recited poems, they had plays and sang songs. It was lots of fun, and a good get together.

— Frena Cutler (VFC excerpt)

Along Main Street

During the 1900s, the houses in Moretown were crammed so close to each other, they were, in some cases, just six feet apart. This increased the fire danger enormously as happened when the upper store burned down.

At the far right is the Town Hall, to its left is the wheelwright shop

Right: on the left of the photo, is the corner shown in the photo below. The corner house still exists, but the tenement house burned down and is now parking lot for the General Store, formerly the Ward IGA Store (the lower store). See the photo on the facing page for today's view of the General Store and parking area/gas station.

All photos in this chapter, save the photo of today's General Store, courtesy of the Moretown Historical Society

Moretown had a band

Above: Moretown's circa 1900 military band had 25 members. This photo is undated.

Right: The band marching through town.

Below: The bandstand built in 1900 hosted the military band. It was located to the right of the current Frank Piazza house.

The Family Through the Years

Burton Smith Ward

When Burton Smith Ward (known locally as B.S.) was born in 1873, his father had only recently started the mill on Dowsville Brook. Within a very few years, Burton saw his father develop that small business into a spectacular success. Soon Burton began his own long tenure directing the company. Of all the Ward men, Burton was the only one who worked with all four generations in the family firm.

After his Duxbury childhood, Burton was a teenager when Hiram and May Ward moved the family to Moretown. He studied at Montpelier Seminary and graduated from St. Johnsbury Academy in 1893—the same year that he and his brother Clinton went to the Columbian Exposition in Chicago.

During his young adult life, the mills were prospering. A boom in industry and construction drove strong markets for lumber. Along with providing high-grade lumber and clapboards, the Ward mills were producing hundreds of thousands of feet of lumber for boxes which used lower grades of lumber. Cardboard packaging was not yet available and wooden boxes were essential for all kinds of shipping. Later, in the 1930s and 1940s as market demands changed, Burton was there for the company's transition to greater hardwood production.

The idea of planting trees and reestablishing forests was in its early years when Burton Ward recognized the importance and value of reforestation. He went on to lead Ward reforestation efforts, overseeing acquisition of lands and planting and pruning well over a million pine and spruce trees. These trees yielded high quality lumber many years later. Burton's vision, hard work, innovation, and enthusiasm earned him recognition as one of Vermont's foremost advocates for reforestation. He was recognized by Vermont Junior College in 1950 for his leadership in forestry.

His enthusiasm for forestry and his business sense are evident in a speech he delivered in 1931 that was subsequently published in *The Vermonter* magazine (pages 75–78).

"It is only within a few years that my eyes have been opened to the grandness and soundness of this undertaking. To be sure, it seems like child's play when we are setting out these little transplants, but what work more is interesting and thrilling than to set out a tiny tree, at the cost of a penny, and see it grow to be worth one dollar, two dollars, ten dollars: yes, twenty, in some cases if given time to mature."

He eloquently asked:

"What project is more worthwhile than that of helping to restore prosperity to our rural communities; the feeders of the larger towns and cities, by growing forests."

Beyond the mills, Burton also devoted himself to agriculture, and took particular pride in his herd of Guernsey cattle. He kept his cattle at two farms in Moretown and another in Fayston. Hay was cut on the old family farm on Ward Hill in Duxbury for many years. Once that land went out of agricultural production, it was planted with trees.

A pillar of the community, Burton Ward was a lifelong member of the Moretown Methodist Church. Every Sunday he was at services with his family, the Wards seated in the front pews. Burton served as a director of the Capital Savings Bank and Trust Co., of Montpelier. He was also involved with lumber industry and dairy organizations.

On an October day in 1951, Burton Ward passed away. He had been out in his beloved tree plantations. As his obituary noted:

> Burton S. Ward, 78, prominent lumberman, died yesterday afternoon at 3:30.
>
> As he did on every day he possibly could, he had taken his pickup truck with some young men to work among the trees of his forests in North Fayston. Their work was completed and Mr. Ward had gotten into his truck to come home when he was suddenly stricken.
>
> Just the previous day he had attended the Forest Festival program in Montpelier.

Burton, Clair and Merlin, 1945

Burton's Family

Annie Laurie Evans — 1895–1955

Burton Smith Ward married Annie Laurie Evans of Moretown on October 17, 1895 in the Moretown Methodist Church. The ceremony was officiated by Burton's uncle, the Reverend Earl Ward of St. Johnsbury, Vermont. Burton and Annie were parents of the next generation to run the mills: Merlin Burton, born October 3, 1897; Kenneth Hiram, born March 1, 1901; and their daughter Marion, born November 5, 1904.

She is fondly remembered for her love of music, her fondness for skating on the river when conditions permitted, and her devotion to her grandchildren. Especially remembered was her exceptionally warm disposition. It is known that following in May's tradition, Annie was a active in the Methodist church. In 1940, at a special Easter evening service, she was honored with a trophy by her friends in the choir for her 40 years of participation. After a long, active life, Annie Evans died on December 1, 1955.

Their children: Kenneth, Marion and Merlin

Courtesy of the Moretown Historical Society

Kenneth, Merlin and Marion Ward, circa 1905

Burton with grandson Richard in 1926

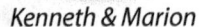

Kenneth & Marion

Annie Evans wrote down this family recipe. It was found in a cookbook years later with this note: "It's in Annie Ward's hand apparently given to Gram Evans…"

Marion Ward and sister-in-law, Aline

Right: Gram Evans, Aline, Marion, Mrs. Annie and Merlin

Kenneth Hiram Ward — 1901–1942

The years of the 1930s brought repeated challenges to the Ward family business, including property losses from floodwaters and fire, and the difficult business climate of the Great Depression. In 1942, great personal loss was added to list. Kenneth Hiram Ward, only 41 years old, died in September of a brain tumor.

Kenneth was born in Moretown in 1901, the same year the family was settling into the gracious new Big House. His formal education started at the Moretown School in the village and continued at Montpelier Seminary. While studying at Dartmouth College he belonged to Phi Gamma Delta fraternity and distinguished himself in athletics.

After graduating from Dartmouth in 1924, he returned home to Moretown, and married Florence Miles of Sheffield that July.

In Moretown, Kenneth joined his father, Burton, and his brother, Merlin, in the lumber business. For the company to run smoothly, these three family members knew all aspects of it—from planting and pruning trees, to logging management, to mill operations, and running the store. Kenneth especially focused his attention on supervising the mills. He had a key role in moving the company forward. In 1935, he made opportunity from adversity. After the loss of the Lower Mill to fire that year, the company built an improved mill on that site. Installed equipment included a band saw that improved efficiency. That modernization helped position the firm for coming years. Another of Kenneth's initiatives was Ward Lumber's acquisition of about 5,000 acres of woodland on Camel's Hump. With much maple and ash, this parcel in the Winooski watershed became a valuable resource.

Kenneth was a director of Associated Industries of Vermont, a statewide organization of manufacturing businesses. In the community, he was active in groups including the Mad River Valley Grange. With a fine voice and a good ear, he sang in the church choir and for fun, often with Florence. He was a lifelong member and steward of the Moretown Methodist Church. An enthusiastic athlete, baseball was his favorite sport. He played on the Moretown baseball team and coached the youth team.

In 1937, *Dartmouth Alumni Magazine* described Kenneth,

> Ken Ward lives year round in the same territory. He is mill superintendent of the Ward Lumber Co., Moretown, Vt., oversees four lumber mills, purchases logs for same, and directs the planting and caring for some 900,000 trees. The first of his two sons, Owen 12 and Wyman 3, unquestionably plays on his father's ball team, the Moretown Sluggers, consisting of boys from eight to fourteen. Sidelines include the masons and trusteeship of the Montpelier Seminary and Vermont Junior College.

Owen, Kenneth and Florence Ward

Kenneth Ward's early death is especially sad because his abilities were tested in such difficult times. New opportunities and technology for the lumber industry were just dawning. His family and community never had the opportunity to see his potential in the coming era.

Dartmouth Alumni Magazine:

Kenneth Hiram Ward died at his home in Moretown, Vt., September 26. He had been in failing health since early in June, and was a patient at the Neurological Institute in Montreal in July and August, where his case was pronounced incurable.

The son of Burton Smith and Annie Laurie (Evans) Ward, he was born in Moretown, March 1, 1901, and prepared for college at Montpelier Seminary. He was a member of Phi Gamma Delta, was captain of the wrestling team, and played baseball and football.

Since graduation he had been in the lumber business with his father and brother, being the vice president of the Ward Lumber Co., and he was also agent for several insurance companies.

He was a steward of the Methodist church, and for many years had sung in its choir. He was a member of the Mad River Valley Grange, a past master of Mad River Lodge, F. and A. M., and a director of the Associated Industries of Vermont.

Owen remembers his father:

"He was a gentle, although powerful, man in physique and in character. He was very personable and his leadership was demonstrated in many fields. He was known for his integrity in business. I still have a letter from one of the pulp companies he dealt with; the letter tells of his integrity and their respect for him.

"For fun, he was instrumental in baseball in Moretown. He helped nurture interest in the game in young players, coaching and encouraging us."

Wyman, Kenneth and Owen at Lake Elmore

Florence Miles Ward — 1900–1958

Kenneth's wife, Florence, was an active partner in the Ward Lumber Company serving as a clerk, and as a director, as well as working for the Town of Moretown as auditor, town clerk, assistant town clerk, treasurer and health officer. She also managed a term in the Vermont State Assembly serving as a representative from Moretown before her illness.

Each year, the State Legislature publishes a listing of all the members of the Assembly. In the *Journal of the House of the State of Vermont, Biennial Session 1957,* her biography and committee assignments read:

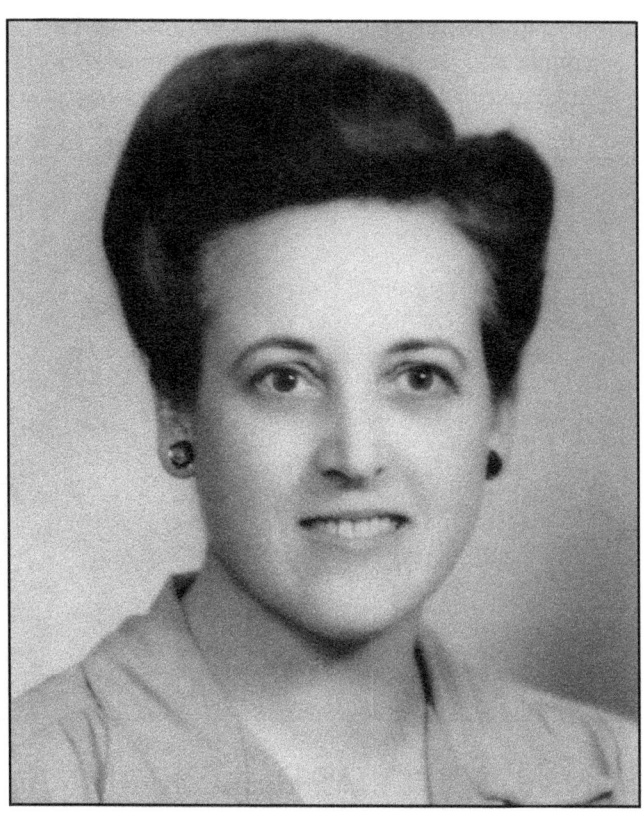

> WARD, Florence Miles (Mrs. Kenneth H.), Representative of Moretown, Republican, was born in Glover on August 30, 1900 and located in town in 1924. Occupation: clerk and director of Ward Lumber Co., Inc.; she was a teacher prior to marriage. Educated in Glover grade school: Barton Academy, 1917; teacher training course, Montpelier Seminary, 1918; Syracuse University, 1921-1923. She is a widow and has two sons, Member P.T.A.; Grange and Farm Bureau. Secretary-treasurer Republican Town Committee and vice, chairman Republican County Committee. She has served the town as auditor, health officer, town clerk and treasurer, and presently is serving as assistant town clerk. Member and permanent historian of the House of 1955. Member commission on Interstate Cooperation.

MRS. FLORENCE M. WARD

Moretown Town Representative Dies Following Long Illness

MORETOWN - Mrs. Florence Miles Ward, 57. Moretown town representative in 1955–1957, died early this morning after a prolonged illness.

She was born in Glover Aug. 30, 1900, daughter of Orrin G. and Mae (Woodward) Miles. She was graduated from Barton Academy in 1917 and from Montpelier Seminary in the teaching course in 1918.

The deceased attended Syracuse University where she was a member of the Delta Gamma Society and taught for three years in Vermont schools, one year in New Hampshire. Her marriage to Kenneth H. Ward of Moretown, took place in 1924 and she had lived in Moretown since that time. Mr. Ward died in 1942.

Surviving are two sons, Owen of Moretown and Wyman, who is in the armed forces; two sisters, Mrs. Mildred Colby of Bath, N.Y., and Mrs. Hazel Jackman from Groveton, N. H. Mrs. Ward was a member of the Moretown Methodist Church which she served as treasurer since 1924 and was a Sunday school teacher and choir member. She was president of the WSCS in which she held other offices.

The deceased was also a member of the Troy Conference board of education, serving as secretary of the Troy Conference student loan and scholarship fund. She was a member and past president of the Moretown PTA and member of the Mad River Valley Grange.

Ward served as clerk and director of the Ward Lumber Co., Inc., since the death of her husband, and represented the town in the General Assembly in 1955–57 where she was a member of the commission on interstate cooperation. In 1957 she was chairman of the committee on Social Security.

Owen remembers his mother:

"My mother was a lovely and very intelligent and capable woman. It is so sad that her years with my father were cut short. They were a wonderful couple and were great parents. They were both musically inclined. She was a pianist and had a beautiful alto voice. She and my father used to sing together—and they were very good.

"She missed him very much, but accomplished so many things in Moretown and in the Legislature in her remaining years."

Kenneth H. Ward Memorial Access Area

Today, local residents, visitors to the area, and fans of the Mad River, enjoy a terrific recreational resource, thanks to Kenneth Ward. His sons, Owen and Wyman, gave the State of Vermont a spectacular gift, donating a piece of land on the Mad River for public use. This property, located along Route 100B, had been part of one of their tree plantations. Kenneth Ward worked on the woodlands on and around this site. Big rounded boulders, deep pools, and a gravel beach make this spot one of the most scenic and popular swimming and fishing holes on the river. Throughout the summer, families, community members and visitors swim, picnic and fish here. The site is officially known as the Kenneth H. Ward Memorial Access. As people enjoy the river and Vermont's natural beauty in this special place, Kenneth Ward's legacy lives on.

Merlin Burton Ward — 1897–1979

In his 81 years, Merlin Ward's life spanned most of the century of the Ward Lumber Company. During his childhood, his grandfather was at the helm of the family business and was passing it on to his son, Burton. A few decades later, Merlin was in that position himself and bringing his son and nephew to its leadership.

Born in 1897, Merlin was Burton and Annie's eldest child. He attended the elementary school in Moretown Village and graduated from Mount Hermon School. During World War I, he served as an ensign in the United States Navy. Merlin Ward's sword, dress uniform, and photograph from his service are now in the collection of the U.S. Naval Museum. After the war, he finished his studies at Syracuse University, graduating in the class of 1920. At Syracuse, he met fellow student Aline Hollopeter of Camden, New Jersey. They married in 1921.

Merlin returned to Moretown after college to work in the family business. The business was comprised of the mills, the plantations, logging on Ward timber lands, and the store. Merlin was involved in all aspects of the company, but had a particular focus on the management of the store. Beginning in late 1923, he added Moretown Postmaster to his duties. He served as postmaster for 39 years. Conveniently, for customers as well as for him, the post office was located at the store; townspeople could pick up their stamps and letters and groceries and necessities in one stop. Adding to the convenience, the lumber company office was right upstairs.

When Kenneth Ward, Merlin's brother, died in 1942, not only did the family suffer a great personal loss, but the company lost a valuable member of its management. Instead of three family members sharing the work and decisions, only Merlin and Burton were left and Burton was already in his advanced years. The fourth generation Wards were still teenagers.

In the 1940s, Moretown, like the rest of the country, dealt with the demands and trials of World War II. Many young Moretown men, including Ward sons, served in the military. Rationing was in effect, affecting industrial operations as well as personal lifestyles. After the war, innovations like chainsaws and new, more powerful mill equipment began taking over from traditional tools and techniques. As the 1950s opened, Merlin was joined in leadership of the company by his nephew, Owen, and his son, Holly.

Merlin was a lifelong member of the Methodist Church. Like many other Wards, he had a fine voice and sang in the choir. For years Merlin tolled the church bell every Sunday morning—his children and nephews occasionally accompanied him in this personal duty. He was a quiet man, fond of reading and writing. He was involved in professional and service organizations. The lumber industry

Merlin in the U.S. Navy (above) and as a cadet with two mates, cadets Peterson (above left) and Bomgartner (below center) in 1917. Peterson was later killed in action.

changed immensely during Merlin Ward's many years. He is remembered for working hard in a challenging business.

Prominent Moretown Figure Merlin B. Ward Dead At 81

MORETOWN - An era in this town came to an end here Wednesday morning with the surprising discovery of the death of Merlin B. Ward.

Ward, husband of former legislator Aline Ward, was a prominent citizen of Moretown for more than a half century, operating the town's 'company' store and serving as town postmaster for 39 years.

Merlin Burton Ward

Merlin Burton Ward, 81, died unexpectedly Wednesday in Moretown.

He was born in Moretown Oct. 3, 1897, son of Burton S. and Annie (Evans) Ward.

He attended local elementary schools, Mount Hermon Preparatory School, and was graduated from Syracuse, University in 1920 with a bachelor of science degree. He was a member of Phi Gamma Delta fraternity.

He was a lifelong member of Moretown United Methodist Church and served as a trustee for many years.

He was a veteran of World War I, having been commissioned an ensign in the U.S. Navy.

With his father and brother Kenneth, he operated the Ward Lumber Co., Inc. and was its president when the mill and real estate was sold in 1969.

He had been a director of Northeastern Lumber Manufacturers Association, and of the Automated Industries of Vermont. He had been a trustee of Central Vermont Hospital, and an incorporator of Northfield Savings Bank.

Mr. Ward operated the company general store in Moretown for many years, and was Moretown postmaster from 1923 to 1962. He was a justice of the peace and a member of the regional planning commission for many years.

He was a member of the Montpelier Kiwanis Club. He was a prominent Mason for over 50 years; a past master of Mad River Lodge No. 77, and past patron of Morning Star Chapter No. 23, Eastern Star.

He had been head of several Masonic bodies, including Waterbury Chapter No. 24, H.A.M., Mount Zion Commandery No.9, of Montpelier, Gamaliel Washburn Lodge of Perfection, Mount Calvary Council Princes of Jerusalem, and Frank J. Martin Chapter of Rose Croix. He was a member of Vermont Consistory of Burlington, Knight of the Red Cross of Constantine, and Mount Sinai Shrine Chanters.

He is survived by his widow, the former Aline Hollopeter of Camden, N.J., whom he married in 1921. She was a classmate at Syracuse University, also graduating in 1920.

Also surviving are two sons, Richard S. Ward of Essex Junction and Holly M. Ward of Waitsfield, a daughter, Lois (Ward) Tierney of Dallas, Texas; also a sister, Marion (Ward) Tweedie of Walton, N.Y.; 12 grandchildren and two great-grandchildren.

Aline and Merlin Ward celebrating 50 years of marriage

Merlin and Aline in 1919

1928–Merlin with Richard and Lois

Aline Hollopeter Ward — 1898–1994

After graduating from Syracuse University, where she met Merlin, Aline taught high school back in her hometown of Camden, New Jersey for a year, then came north to Moretown. Merlin and Aline married on June 28, 1921 in Camden returning to Vermont in August.

Aline was elected to the Vermont House and later Senate. She was an active member of the local school boards from 1936 to 1966 up to and including the building of Harwood Union High School.

As a member of the Vermont State Assembly, she was responsible for passing much legislation. Her *Journal of the House of the State of Vermont* biography reads:

Aline's 25th wedding anniversary photo

Aline H. Ward (Mrs. Merlin Burton) of Moretown, Republican, was born in Camden, New Jersey on June 17, 1898, and became resident of present town in August, 1921. Occupation: (former high school teacher; part-time secretary, Ward Lumber Company, Inc. of Waterbury. She was educated in Camden, New Jersey schools; and Syracuse University (A.B., 1920), and one summer of graduate work, University of Vermont. She is married and has two sons and one daughter. She is a member of Washington County Farm Bureau (director); town school board (director); and Vermont College (trustee). She has been a member of Vermont State School Director's Association, 1962 (president); Vermont Congress P.T.A., 1953–56 (president); Vermont Civil Defense, Women's Organization, 1958–61 (chairman); and Unemployment Compensation Committee, 1957–61. Member of the Senate of 1961, 1963. Member of the House of 1947, 1953. Religious preference: Methodist.

Aline H. Ward

Aline H. Ward, 96, of Moretown passed away at the Webster Home in Rye, New Hampshire, on Monday, June 27, 1994. Born in Camden, New Jersey on June 17, 1898, she was the daughter of John and Margaret (Smith) Hollopeter. In 1921 she married Merlin B. Ward, a fellow graduate of Syracuse University and past president of Ward Lumber Co. of Moretown. Her husband died March 7, 1979.

She was a 1920 graduate of Syracuse University where she was a member of Delta Gamma sorority. She taught English at Camden (New Jersey) High School from 1920–1921. An active member of her community, she served as Moretown's representative to the legislature for three terms and as a Washington County State Senator for two terms. She was a delegate to the Republican National Convention in 1956, when Dwight D. Eisenhower was nominated for president.

She also served as president of the Vermont Federation of Republican Women from 1969–1971. She served as a trustee for Norwich University from 1971–1978 and was awarded an honorary doctorate degree in Humanities in 1973. Her other memberships included the Queen Esther Chapter #7, OES, and was a lifelong member and loyal supporter of the Moretown United Methodist Church.

Aline receiving an honorary degree from Norwich University.

Reforestation circa 1910 to 1968

"Timber for Tomorrow, Too," was the title of a feature article published in the Autumn, 1962, issue of *Vermont Life* magazine. Penned by acclaimed naturalist and writer Ronald Rood, the piece opened with the question, "What happens to the Vermont woodland after you cut the trees?" Rood spent time in the Ward mills and talked extensively with Merlin Ward, then company president, for the article. The story focused on the Wards' long and distinguished practice of reforestation and maintenance of their woodlands.

> One of the wood-users who can give a report is the Ward Lumber Company, Inc., of Moretown. Merlin Ward, president, or his nephew Owen, vice-president, or his son, Holly, clerk of the corporation, can predict the fate of nearly every acre of woods they cut for the next ten, twenty or fifty years. And in a state nearly two-thirds wooded, knowing the probable future of a patch of forest can be mighty important—especially since about one family in three depends on forest products for a living.

While Vermont was about two-thirds forested and one-third clear at the time Ron Rood wrote about the Wards' practice, the opposite proportion was in place when they started their forest management. Photographs of Vermont of the early 1900s are reminders of that time when most of the state was cleared. Vast expanses of meadows and pastures, not forests, defined the landscape. The state's forests had been leveled to make room for farms. In 1880, Vermont reached its all time maximum number of farms—35,522

Vermont Life *article by Ronald Rood, Autumn, 1962*

in all. Soon after, though, "farm abandonment" became a concern in many rural areas. It was hard to make a living from a small hill farm with thin soils and steep slopes. In the Midwest, a region accessible thanks to the railroads, land was flatter and soil deeper; farmers there could produce wheat and other crops at lower costs. Vermonters were also moving to new opportunities—to cities with thriving industries or to the growing western states. Farmland, when abandoned will eventually grow into forests again—but the first growth trees are of species that are not of much value for lumber.

In the early 1900s Hiram and Burton Ward considered their future and how to supply logs to keep their mills running for later generations. Newspaper articles show that Hiram Ward owned a substantial amount of land by then—"large tracts of woodland in Dowsville and on Moretown mountain, enough to supply his mills for many years to come," according to the August, 1893, *The Vermont Watchman*. With farmers moving off the land, a lot of acreage came up for sale. Farms in some parts of the state that sold for $100 to $200 an acre in 1874 sold for less than

Abandoned farmhouse in Fayston

$10 per acre at the end of the century. The total valuation of Vermont farms dropped from $135 million in 1870 to $100 million in 1890.[1] Fallow farmland was not much good for supplying logs in the short term—but for businessmen with a longer view, it provided opportunity.

The decline of local forests and timber supplies concerned the Wards and other Vermonters. In 1904, for the first time, the Vermont legislature authorized the Governor to appoint a Forest Commissioner. Commissioner Arthur Vaughn made these recommendations to the legislature in 1906:

1. We should secure accurate information as to our present forest condition.
2. It will be observed that about half of the fires of last year were caused by railroads. Legislation to lessen this danger seems to be called for.
3. Our great obstacle to the planting of waste lands at present is the excessive cost of little trees. If the State would cooperate with the Agricultural College in the establishment of a forest nursery (a beginning has already been made by the College) material for planting could be furnished our people at less than half present cost.
4. Perhaps the matter of most pressing importance, the present legislature should not adjourn without making adequate provision to fight any invasion of gypsy moth and long tail moth.

Following these recommendations, the Vermont State Tree Nursery was soon established. This nursery, as an affordable source of seedling trees, proved to be a terrific resource to the state's wood products industries.

The 1962 *Vermont Life* article by Ronald Rood, illuminates the beginnings of the Ward Lumber Company's reforestation projects. "Much of the credit for the company's present status goes to Burton Ward, son of the company's founder," Rood wrote. "When he and his father met Gifford Pinchot, the great forester, about 1912, they were impressed with the possibilities of reforestation. They began buying abandoned farms, holding the timber on them and planting the fields to spruce and pine." Burton Ward led and sustained this practice for decades and passed it down to the next generations.

Burton Ward was widely recognized as a leader of Vermont reforestation. Forestry programs, including those at Yale University and Syracuse University brought students to Ward lands to see his practices at work. In 1931, he spoke at a

---

1 Jan Albers, *Hands on the Land*, p. 204

forestry meeting about the value and importance of reforestation. His speech was published in the October, 1931 issue of *The Vermonter,* a state magazine that pre-dated *Vermont Life.* His words are still compelling today.

Vermont Forestry

by Burton S. Ward

THE rapid depletion of the forests of this country by the ax, fire, and disease has progressed to a point that it seems alarming to those who take a long view into the future. On many of the farms in our own state, there is not a single spruce tree left standing that is fifteen inches on the stump and practically none growing. The old maples are also disappearing and not much is being done to encourage the second growth.

The Federal Government and the States, recognizing these conditions are trying to develop a policy that will tend to conserve our timber by securing more intelligent cutting and stopping waste by fire and disease; also by encouraging reforestation on cut-over, abandoned, and burned lands.

The progress to date has been as good as might be expected when we consider the nature of the undertaking; namely to get men to invest money in a crop to be harvested after they are dead.

To many this expenditure of time and money may not seem necessary; as our own hills and mountains look quite green as we view them from a distance. One might think, from a casual glance, that too much of our land is growing up to timber; but what are the facts?

If we will take time to examine this new growth we will find a large percentage of weed trees; such as Alder, Willow, Pin-cherry, Gray Birch, Poplar, Elm, and I might almost include Hemlock and Beech, when we want Maple, Ash, Birch, Bass, Spruce and Pine.

We all know it is not only necessary to plant our gardens in the spring to get a good crop, but also we know it is imperative to weed them if we wish to say what that crop shall be. Just so with our beautiful hills and mountains; if we want to assure ourselves that their covering shall have a commercial value 50 years hence we must aid nature by weeding and pruning.

The question now arises, what land ought to be planted and when?

It is generally thought that the best results are obtained on such open land and abandoned farms as are unsuitable or unprofitable to till because of location, rocks, or broken by hills and gullies. There are many thousand acres of this class of land available at low prices.

I believe the early planting of such areas is of vital importance. It is not appreciated by the majority of our citizens or there would be more planting done each year.

The ideal time to plant is the first year after the land is abandoned, that is, not to be mowed or pastured any more, thus giving the young trees a chance to get started before the coming brush gets large enough to choke and smother the plants.

These open areas set out to transplants ought to be worth $100 per acre in forty years. If nothing is done they will be worth from $25 to $50 less than nothing for they will eventually be seeded in with scrub trees which will cost $25 or more an acre to remove, thus making the land a liability to the owner instead of an asset.

Judging from past experiences some of this land, if not planted, will be as lost to civilization as though it disappeared from the earth.

Many acres of our state which have been tilled, in the last fifty years, are not needed now to grow crops owing to the surplus of better farm lands elsewhere, not to mention the large tracts which may become productive through irrigation.

If in years to come this land should be needed for tillage it can be redeemed. There is certainly time enough to grow one crop of trees before it will be needed for agriculture. As the soil is ideal for planting, trees will grow rapidly.

Vermont's soil is surpassed by no other New England State and equaled by only one.

Along Ward Hill today, beyond the greenery in front are trees planted generations ago by the Ward Lumber Company. This location is across the road from H.O. Ward's original home.

Recently a Forester, connected with the Forestry Department of Syracuse University, who has a nursery in Massachusetts, inspected some of our plantations. He was exceedingly enthusiastic over the growth the trees were making and their near freedom from the weevil. He said that the growth was equal if not superior to anything he had seen anywhere. Vermont soil fed those trees. Give Vermont the credit.

It is only within a few years that my eyes have been opened to the grandness and soundness of this undertaking. To be sure it seems like child's play when we are setting out these little transplants, but what work is more interesting and thrilling than to set out a tiny tree, at the cost of a penny, and see it grow to be worth one dollar, two dollars, ten dollars: yes, twenty, in some cases if given time to mature.

I well remember one field that we planted where there was much wild grass and heavy sod. From time to time, during the next two years, I looked for trees on that plot and could see nothing but grass and weeds and decided it was a 100% failure. Some time later, after the frosts had killed the grass and weeds and there was about two inches of snow on the ground, I came in sight of that field and although I was over a mile away I could see the green heads of thousands of these trees, the Forests of Tomorrow, sticking up through the snow.

What project is more worthwhile than that of helping to restore prosperity to our rural communities; the feeders of the larger towns and cities, by growing forests. This gives work to the farmers in slack times, bedding for his cattle and labor for the villagers.

What patience should we have with the man who says it is foolish to spend money on our forests when we can get any quantity of lumber from Canada, the Pacific Coast States, the Southern States, and Russia.

We take interest in and are proud of our milk and butter, our maple products, our apples, our granite and marble, and talk about roads much of the time. Why not talk Forests a while and show a little of the enthusiasm for them that we do for these other things.

Let us teach our children Forestry in the grammar and high schools and speed on their work in our universities.

Thus through the education of the boys and girls we may hope to so interest them in forestry that when the responsibility of government shall rest upon their shoulders they will see that appropriations commensurate with its importance shall be voted by our legislature.

The present appropriations are very small and in some cases the lumbermen themselves are partly responsible for this lack of interest and the apparent prejudice which exists.

The portable mill has often done irreparable injury to a town by stripping its mountains of timber, using transient help to clean their holdings and then move on; thus acting the part of the vandal instead of a community builder.

I sincerely hope their day is about over and that in their place we have the resident millman who takes a vital interest in his community and by selective cutting, combined with reforestation, assures that community of a perpetual industry.

Fully realizing our responsibility I trust that we shall so operate that we shall be looked on as builders of the state and not destroyers.

Forestry's chief support comes from progressive, broad-minded men and women who can visualize the results of their efforts fifty years hence, men and women who do not dodge responsibility but are glad to do their part knowing that what this country has to offer for their children and grandchildren depends in large measure on what they do to develop and conserve its natural resources.

The editor of *The Vermonter* added a short profile of Burton Ward, recognizing his character and his achievement.

Mr. Ward was born in Duxbury, Vt., Feb. 13, 1873, living there until 1889 when, with his parents, he moved to Moretown and has since resided there, succeeding his father, H.O. Ward, doing business under the name of Ward Lumber Co. They manufacture

Owen in a plantation forest, April 13, 1949

clapboards, dimension stock and hard and soft wood lumber in the "upper and lower mills" at opposite ends of the village, one of which has a tremendous potential water power, if developed, with natural rock abutments. The Company also has an electrically operated mill at Middlesex. The lumber is sold in cities of New England and New York.

Mr. Ward is a staunch believer in Vermont, as evinced by the accompanying article, which was delivered at a forestry meeting. He owns several farms, has large areas of timberland in Duxbury, Fayston, Warren, Waitsfield, Northfield and Moretown, and has a number of "marginal farms" on which in the last 20 years he has planted over 600,000 trees, more than any other individual in the state. It is against the policy of the company to post such lands against hunters and fishermen. All such are welcome if careful about fires.

Mr. Ward has two sons, both college graduates, who are in business with him. Merlin B. is general store manager and postmaster, while Kenneth H. is mill superintendent. A married daughter. Marion

L., a graduate of Syracuse, is supt, of art in the Walton, N.Y. schools.

His principal hobby is the breeding of registered Guernsey cattle, 80 head. The Mad River Fox Ranch is also an interesting venture of the Company.

The village of Moretown is in the southwest corner of the township, along the Mad River. The other corners are close to Waterbury, Montpelier and Northfield.

The interests of Mr. Ward are closely allied with those of the village of Moretown. During the recent depression the mills have been kept running by rigid economy and close attention to costs, thus employment has been furnished to practically all the workers of the village and for several from adjoining towns. Living costs are very reasonable as Mr. Ward furnishes many with homes and fuel at about one-half the price asked in larger communities.

Good fortune seems to have attended the people of Moretown. A splendid new Superior Grammar School now adorns the village and new cement walks, paid for by abutters and town jointly, just completed this fall, extend nearly the length of Main street, on both sides.

Mr. Ward is much interested in state, county and town affairs. He is consistent in attendance upon divine service and he is another of those sterling souls who like to get back, with the family, and drive along so-called back roads, away from the rush of automobile traffic.

Men of Mr. Ward's type, who stay by the home town and defy tradition, who survive adversity (the firm's flood loss was $75,000) who circumvent depression, add to the prosperity of a region and look hopefully forward, mean more to Vermont than heroes of war or adventure. We cannot honor them sufficiently.

C.R.C.

The Ward children all did their part in the plantations.

Lois Ward Tierney recalled that her grandfather, Burton, had a tiny pruning saw made for her when she was a little girl. She would accompany her grandfather to the plantations and learned to use her saw at a young age.

An interview with Cedric Reagan:

We'd get paid one cent per tree to prune: got up to a dollar a day and got sick of it when it was a hot day. Burton would follow and count the trees in our rows. We'd get a slip of paper and take it to the office to get paid.

Robert Wimble pruned. Ozzie too.

Sometimes some of the older girls would prune and they were good workers and would outdo us boys. Jean Hurdle pruned. (Her mother worked in the box shop.

—In a 2011 interview with a reporter

Planting and Pruning

Planting trees for future needs requires a long view. As Burton Ward observed in 1931, "The progress to date has been as good as can be expected when we consider the nature of the undertaking; namely to get men to invest money in a crop to be harvested after they are dead." Four generations of Wards took that long view. Trees that they planted during World War I, became pianos, furniture, and houses of the 1950s and 1960s. Trees planted during the Great Depression became lumber of the 1980s. Some trees planted in the company's later years are being cut and milled in the 21st century. Totaling various accounts, the sum of Ward planted trees may have surpassed two million.

Burton Ward was a champion of Vermont forestry. A solid Yankee aversion to waste likely contributed to his enthusiasm for tree plantations—in his essay he commented on the wasted land of abandoned farms. His focus, though, was his knowledge of the lumber industry's need for timber. Tree plantations, given enough time, would provide logs for the company. Well-managed tree plantations would provide logs that would yield high-quality lumber—lumber with fewer flaws and thus greater value than that of many naturally growing trees.

Land, planning, seedlings, labor, and time were the main ingredients of tree farms. Bog hoes and pruning saws were the tools. The principal species of the Ward plantations were white pine and spruce, although Burton also planted some red pine and experimented with some other species. White pine and spruce, softwoods, are indigenous to the region. They have thrived in Vermont for millennia—since forests began growing again after the glaciers receded 10,000 years ago.

White pine and spruce grow quickly, once established. A mature white pine, after its first 10 years or so, may grow one-half inch on a ring in a year—translating to an inch of diameter growth. A sugar maple or other Vermont hardwood may only grow $1/16$ of an inch on a ring, a mere $1/8$ of an inch in diameter during the same year. Observing the cross section of a log, it is evident that growth varies—some rings are closer together, some are farther apart. Every year, a living tree grows in circumference, growing new layers of

Vermont Department of Forests Parks and Recreation, Forestry Centennial

1911, planting Scotch pine in Plainfield. "Snowed during the planting. Within two weeks intense hot weather came on and killed about ½ the trees."

cells. Larger cells with thin cell walls develop in the warm weather of spring and summer. These appear as lighter colored wood. Cells grown later in the year—in fall and winter—are smaller, have thicker cell walls, and appear dark in color. Environmental conditions including water and temperature influence a tree's growth.

In 1906 or 1907, the State of Vermont established a tree nursery at the University of Vermont. The purpose of this nursery at Centennial Field was to provide an affordable source of seedlings for reforestation. In 1922, the nursery was moved to Essex, where it remained for many years. Seedlings were also available from commercial suppliers.

From the time Burton and Hiram Ward first started the tree plantations, apparently around 1910, Ward Lumber Company planted almost every year, as long as open fields were available. Planting was generally done during slow times at the mill. Employees who would otherwise be working at mill jobs instead went out in the fields. They followed a time tested routine for starting the little trees.

"It was arduous labor," recalls Owen Ward. The young men of the Ward family also put in many hours planting seedlings. Owen recalls how Clyde Baird, Burton's right-hand man, often directed the planting.

"Clyde Baird was amazing. He could walk a dead straight line across a field," Owen recounts. Baird would start at the lower end of a field, laying a baseline for the planting. He had a knack for staying on track even as the terrain sloped up and down. As Baird walked he dug his heel in at regular intervals to show where each tree should be set. Behind Baird, the next man wielded a bog hoe. Two sharp strikes to the ground with the pick end of the tool laid deep parallel cuts through the sod and root. A third strike with the flat end of the hoe connected them creating a flap of sod.

"The guy with a pick would make a mark on the right hand side of the sod and the left hand side of the sod. Then he'd turn it over and take that flat part of the bog hoe and drive it into the ground and pull back a flap. He would pull that flap back. The guy behind him was carrying trees. They might be only 10–12 inches, 15 at the most. He'd put the roots of the sapling in the hole, then take his foot and stamp that flap back in. Then he'd go to the next one, and so on all the way back and forth across that field...."

"We planted most of them six feet or eight feet apart," explained Owen. Burton experimented with the spacing of the trees. The goal was for the trees' growth to go into height rather than into branches. "When they first started planting trees, they planted them six feet apart and then they started planting them eight feet apart because six feet were so crowded on each other you didn't get the growth that you wanted to get. The eight feet seemed to work a little better." Planted at six-foot intervals, a one acre field could hold 1,200 trees.

In one of Burton Ward's experiments, Owen explained, "They tried alternating pine and spruce because they had a weevil that was eating the pine. They figured if they planted both, it would interrupt the flow of the disease from one tree to another." The weevil "caused blister rust

This is the heart of the log, with a clapboard still clinging to it, after most of the other clapboards have been snapped off after sawing. It is easy to see the knot on the left center of the heart, and the remains of another on the top right of the log.

on the pine trees." The alternating experiment didn't work. "What really happened was the pine would grow a lot faster than the spruce." Shaded out, the spruce did not thrive.

Two pests, white pine weevil, *Pissodes strobe*, and white pine blister rust, *Cronartium ribicola*, took a toll on white pine forests of the northeast in the early 20th century. A native insect, the weevil attacks the new growth of a young tree's main stem. Its larvae live under the bark, feeding on the new growth. The tell-tale sign of its presence is a dead main stem turned down like a shepherd's hook. This was a terrible sign for the trees' owners as it meant that the trees were useless for lumber. Blister rust, a non-native fungus, girdles and kills branches and then moves on to attack a tree's main stem. This fungus must spend part of its life cycle on other hosts, currant or gooseberry bushes, to survive. Part of the management of Ward lands included looking for and pulling out these bushes. Removing current and gooseberry bushes limited the spread of blister rust.

Ward Lumber occasionally incurred significant loss of trees from these pests and other problems. Burton Ward was resourceful in trying techniques to limit and avoid losses. Horticulture and forestry students visited the Ward plantations to learn about these efforts as well as about other forestry practices there.

As Burton Ward described in *The Vermonter* article, and Owen and Holly Ward and others witnessed, the first year a field was planted with saplings, its appearance did not betray its future—but within a few years, the young trees burst forth.

The second critical piece of Ward tree plantation management began when the saplings were a few years old. Owen Ward recalls that sometimes Burton initiated pruning when the young trees were three or four feet high. Generally, the trees were pruned twice—once when they were fairly small, again five or ten years later when they were approximately 20 feet tall.

The purpose of the pruning was to remove branches from the lower part of the trunk. The lowest 16 to 18 feet of the trunk provide the timber for marketable lumber. Pruning lower branches encourages vertical growth of the tree. It also improves the quality of the wood.

When a living branch is pruned off of a spruce or white pine, as the tree continues its growth, the annual ring develops smoothly over the place where the branch was removed. In following years, the rings will continue to develop in their concentric pattern.

If a branch is not removed, that ring pattern is disrupted. When the branch dies and is still jutting out from the trunk, the tree trunk grows around that dead branch. Each year, the tree continues its growth. Each ring reaches an obstacle at the dead branch. Over the years, rings build up and the branch is enclosed. When that log goes to a mill, the dead branch is revealed as a knot when lumber is sawed. Knots have a different color, density, and texture than the rest of the wood, lowering the grade of lumber and its value.

Clapboards were central to H.O. Ward's original business and continued to be manufactured by the company through the decades. Vertical-sawn clapboards, like those sawn by the Wards, typically break at the point of the knot which reduces the length of the marketable clapboard and diminishes its value.

Clear lumber has far greater value than its knottier relations. The route to clear lumber was through pruning. Many people in the Moretown and Mad River Valley community had a hand in pruning Ward plantation trees. Mill employees sometimes worked in the plantations, but the company also hired part-time help for this task, including local young people. Pruners were generally paid per tree.

Pruning was usually done in the fall, as there is less pitch in the trees then. Crews of two, three, or four people would head out to the plantations, armed with pruning saws. They carried kerosene-soaked rags for wiping the pitch from the blade to keep the blades cutting smoothly.

"A pruning saw cuts one way—toward you. It doesn't cut when you push it away, so don't waste your effort," said Owen Ward about this familiar task. In general, trees were pruned to the height a person could reach. Occasionally, but only occasionally, they used ladders to prune higher.

Ward Lumber Company pruning continued into the 1960s. The clear lumber of trees cut from those plantations in the 21st century attest to the valuable work of the pruners.

Holly Ward showing the spots in a white pine where the budding branches were pruned back years before

These healed areas on the white pine indicate where branches have broken off, as well as indicating knots in the tree making it unsuitable for clapboards.

Logging & Timberlands

A white pine crashing to the forest floor is a formidable sight—and sound. A mature white pine or sugar maple weighs over a ton. When its vertical stance is undone by a logger's axe and saw, a tree succumbs to gravity. The treetop sweeps out an arc on its earthward fall. Branches snap as they collide with neighboring trees. A tree that stood 60 or 70 or more feet tall hits the ground within seconds; anything in its path is in peril.

Logging is and has always been heavy and dangerous work. Trees, of course, are the foundation of the lumber industry. The first step in the process of transforming these natural resources to manufactured products begins with the calloused hands, strong backs, and sound judgment of loggers.

For many decades, logging was vitally important in the Mad River Valley economy. Many mills, the Wards' included, dotted the banks of the river and tributaries. Rolling pins, butter tubs, packing boxes, clothespins, bobbins, coffins, furniture, as well as clapboards, shingles, and millions of feet of dimension lumber came from local trees milled in the watershed. Ward Lumber became the largest and longest-lived of the local mills, but every mill depended on availability of logs and the productivity of loggers.

The connection between local farms and mills was strong. Many farmers spent winter months harvesting trees from their own forest land. Sale of logs provided cash income—helpful for paying taxes or meeting other needs.

Originally, many of Vermont's sawmills, including Hiram Ward's Dowsville mill, were on small tributaries. In these upland locations, the mills were close to or in the forests. Logs did not have to be hauled a great distance. Hiram Ward's 1874 journal documents the seasonal rhythm of the lumber business of that era. When the Ward mill moved to Moretown and expanded productivity, its need for timber increased.

Hiram Ward had foreseen the value of owning forest land to assure a supply of trees for his mills by acquiring thousands of tree-filled acres. This practice of purchasing productive land continued

A massive white pine about 80–90 feet tall and 100 years old

through the next generations and was supplemented by acquisitions of abandoned farms and establishment of tree plantations. To a large extent, Ward Lumber had its own sources of logs. Through the years the company also regularly purchased from farmers and other landowners and loggers.

Vermont Rule

In the lumber business, quantities are measured in feet. A board foot is a volume of wood one foot wide, one foot long, and one inch deep. The board feet in a log was calculated using "Vermont Rule."

Vermont Rule calculations are based on a 12-foot long log. For a longer or shorter log, the calculation is multiplied by the length's fraction of 12 feet. Vermont Rule is derived from the dimension at the small end of the log measured from just inside the bark line. The diameter is then multiplied by half the diameter (the radius). For example, a 12-foot log, with a 20-inch diameter is calculated: 20 (diameter) x 10 (radius) x 1 (because it is 12 feet long) = 200 feet.

Already marked 480 using the Vermont Rule of measurement, this log is loaded onto a steel-rail skid, and strapped down with a chain.

The Wards visiting a logging site from left to right: Holly, a friend, Owen (in back), Burton, Florence, and Merlin.

In the Woods

Cutting trees and getting them to the mill was a labor intensive process. Axes, crosscut saws, and peaveys were the time-tested tools of the loggers. The traditional ways of logging continued until almost 1950, when chainsaws finally became lightweight and reliable enough to be labor-savers. In the pre-chainsaw technique, the logger used a sharp axe to cut a notch low on a tree trunk. Two men usually worked together. After the initial wedge was cut, they manned the handles on opposite ends of a long crosscut saw. As they alternated pulling, the blade's teeth tore through the wood. The men always aimed their notch and cut for the tree to fall clear to the ground. Without a clear fall, a tree could hang up in the branches of another, dangling overhead. These suspended trees were known as "widow-makers." Fatalities were not uncommon among loggers.

Once a tree was on the ground, the woodsmen trimmed off the limbs and top, so the log had its maximum usable length. Moving logs in the woods, and in every step of their handling, the men used the peavey—a simple and remarkably functional tool. The peavey, related to the cant hook, was named for Joseph Peavey, the blacksmith credited with devising it around 1850. The tool is comprised of a long handle with a spike at one end—the business end. A hook attached a short distance up the handle can only travel up and down, rather than rotating any which way on the earlier tool, swinging open to latch onto logs of different diameters. The handle functions as a lever, allowing a logger to shift and roll even crushingly heavy logs.

Horses, and occasionally oxen, did the heavy work of hauling logs through the woods and to the mills. Long after automobiles were on local roads, horses were still widely used in logging. They could get into narrow places and were adept on rough terrain. In winter, horses could skid logs over the snow and frozen ground with simple sleds. Long traverse sleds with steel runners were mainstays of

Oxen were also used to pull logs

Logs at the staging area, ready to go to the mill

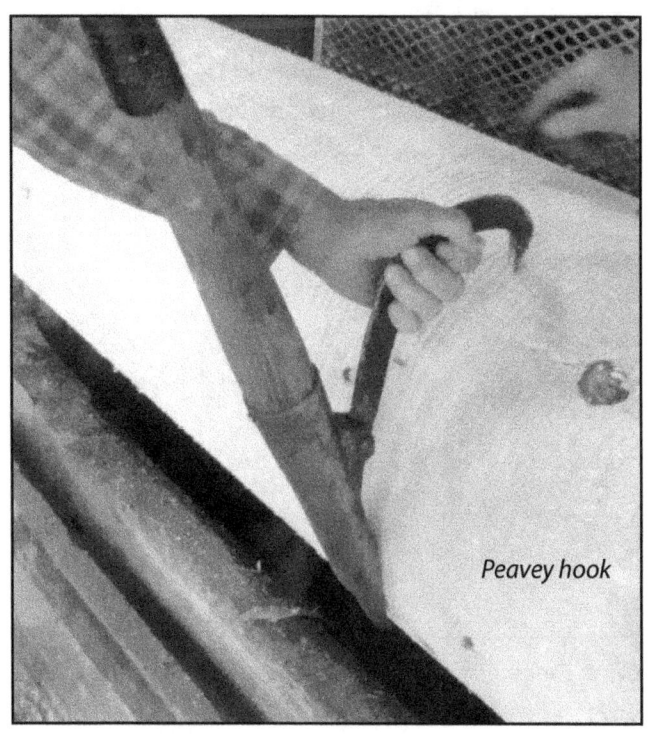
Peavey hook

Peavey bars were used to maneuver the logs up the thin logs used as ramps onto the skid.

logging projects. Ward Lumber Company worked with horses into the 1940s.

Logging in the Green Mountains was usually on hillsides. Logs, once felled, were taken to landings where they were stockpiled until they were hauled to the mill. The trips to the landing and the mill were usually downhill, often down a steep hill. Heavy logs loaded on a sled running downhill quickly gain momentum—a wild ride for the logger on board and a dangerous one for the horses.

Once loaded, one of the men rode on top to make sure the logs and horses made it safely to their destination.

Owen Ward remembers the reputation of one Fayston slope. "There's a hill up the road to Mad River Glen, toward the bottom was a place they called 'Hemlock Hill,' I know they had a big camp up there. Hemlock Hill wrecked a lot of teams because it was so steep and there was a curve that made it very, very difficult."

Large tracts of Ward forest land were in remote areas such as Phenn and Big Basins in Fayston and the uplands of Duxbury. With dirt or mud roads, depending on the season, access was limited. In the pre-commuter days, loggers did not travel daily to and from job sites. Instead, Ward Lumber, as did many lumber companies, set up logging camps. The camps were generally used in the winter as men, many were farmers, were available then. Snow and frozen ground made skidding easier, too. Logging camps included bunk-style accommodations for the men, a cookhouse for their meals, sheds for sheltering the horses and sharpening and repairing tools.

By the 1960s, logging techniques and practices changed with new equipment, forest planning and management.

"We wanted to develop a forest management plan," said Owen, "to guide selective cutting." That plan would include diameter limits for cutting hardwoods and softwoods, as well as guidelines for removal of defective timber.

Leo Laferriere was hired by Ward Lumber in 1964, first as a part-time forester to create exactly that plan.

"I went to work for Wards and my primary area of responsibility was timber harvesting," said Leo. "I would inventory timber, mark sales using maps and report what I found to Owen. I'd bring the company crews into the woodlots and supervise the operations. The company crews were not on the company payroll, they were paid on a production basis; they were paid per thousand board feet."

The Company's truckers and independent truckers including Owen Wimble, went from one job to another to pick up the logs and deliver them to the mill.

Leo remembers, "The crews were kind of varied in number. There was a fellow named Cedric Dunbar, from North Fayston, close to the top of Dunbar Hill. He worked by himself with a crawler tractor. Crews were a maximum of three people. I'll tell you who some of the others were: Eldon DeLong of South Duxbury, he would work either by himself or with another helper. Sometimes they would hire younger people as cutters, but these guys all drove tractors. Eldon had a love for International Harvester equipment. The other crews would characterize him as 'the one driving the red tractor,' Internationals were all painted red. The others used the yellow and green John Deeres. There was Cedric Reagan, who often worked by himself, but sometimes with a helper. Mervin Cutler sometimes worked with Cedric; sometimes he worked with Claude Bonnette. Bunker Wimble sometimes worked with Mervin, and maybe even with Claude.

A sketch hanging in the offices of the Ward Clapboard Mill

"It was Claude who brought in the first skidder. They outperformed the tractor in that they were faster and higher off the ground. They were good for building and leveling the roads into the woodlots as well.

"All the crews were good. They were reliable. But all the men were middle-aged or nearly so, and getting to the point where work was very demanding, physically. After the company was sold [in 1968], most of these men moved on to other lines of work."

Owen added, "With the change to skidders and trucks, the loggers were more productive and they didn't have to feed the horses on the weekends."

Frank Farnsworth stands by his truck, ready to drive the finished products loaded for a trip to the railroad in Middlesex.

> Most [work] was done with horses. There was no hard surface road through the village. And in the spring of the year, people talk about mud on the hills and on the back roads. We had it right through the village. People would get stuck right in the village bringing their logs to the mills. The horses would be really working.
> — Frena Cutler

A Dahlia train working its way through the Warren woods

Vermont Folklife Center interviews
Memories of logging

In the 1992 *Mad River Valley Vermont Folklife Center Project* some residents recalled the era of the logging camps. Many of the interviewees had tales of their own logging days and experiences.

Gerard Dunbar remembered Ward lumber camps from his youth. He and his wife referred to the location of one camp as French Basin on Frenchman Brook. Mr. Dunbar also recalled visiting a second Ward camp.

"I know as kids we used to go up there with a traverse sled, and Carl Richards was the cook, and he told us to come out and eat," Dunbar recalled. The cook, "gave us a good feed." After their meal, the boys, "come down on that traverse sled, and they come like the devil; down through the roads were well packed, and quite steep…. And there was a bridge up there, around the corner, I thought I would go right over that bridge into the brook. Of course the brook was froze over but, there was quite jump there. We made it clear, hooked up on it just once."

Bob Gove, born in 1915, told about his own logging days. He recalled that his brother-in-law, Albert Neill, had a great deal of experience working with big spruce trees, old growth trees, and that those big spruces would sometimes hang up in another tree. Bob Gove explained that Albert, "would take his axe and go right up that tree to wherever it was hung and cut that off, and he would just come down with that tree… just before it hit the ground he would jump off in the snow with his sharp axe in his hand…." Gove explained that he personally opted for a safer approach, "I would hitch my team on to the butt end of that log with a rolling hitch so it would roll when I started my horses, and usually I could roll it out…."

View from the back of the horse pulling a loaded skid

Gove also remembered that his brother-in-law had two teams of horses. "Albert went to work for Ward Lumber Company with those two teams and Pearly Fuller drove one of the teams and he had, he told me, that he had come out of North Fayston, sometimes with two thousand feet of hardwood lumber, logs on a set of sleds with four horses on it, with his false pole to help hold back down the hills and he said there was one hill, that was steep enough so the horses couldn't stop that load, entirely, and he got it, would get his runner chain ready and he'd step right down on the rolls of the sled, with the sleds going, and his hired man would just hold the reins while he was down there and he would have that chain hooked together so that he could drop it over the nose of that sled and he said he would bend over and drop that runner

A wagonload of logs coming through Waitsfield

chain over the nose of the sled to help hold down that steep hill. He said he couldn't stop, he couldn't stop to do it, he had to do it on the go."

Nelson Patch, born in 1909, started logging when he was about 15 years old. His parents had died and he needed to support himself. Patch and his brother worked together for many years, they had their own small camp in Fayston. His brother and sister-in-law also stayed at the camp through the winters. For their labor, in that earlier time, "you got two dollars and a half, a thousand for cutting, you cut a thousand feet and you get two dollars and a half."

For hauling logs, Patch preferred a type of sled known as a "go-devil" over the traditional traverse sled. The loosely constructed go-devils could turn in small spaces and worked well for hauling logs short distances.

Patch also recalled the Wards' impressive horsepower. "And I have seen thirty-five span of horses come right up through Waitsfield and go up to South Fayston and get logs for Ward Lumber Company."

Clesson Eurich, of Waitsfield, born in 1918, explained the local economics of logging, "The farmers used to get out logs every winter to help pay their taxes, that's what they always [did]." On the Eurich family's farm, as on those of many others in the area, they had a woodlot and sugarbush. "I worked for six, seven years I guess for Claude Bonnette for the Ward Lumber Company cutting the big basin back in here."

"Ward Lumber Company—my father used to work in the winter for them lots of times," said Eurich. "They used to haul out of Fayston underneath Burnt Rock, and they'd cut the logs, they'd pile them up and then they'd draw them out in the winter, with sleds and horses clear from there to Moretown to the mill. A lot of the farmers used to work doing that in the winter. The farmers, they did everything to get money. Grampy used to say nobody's going to bed hungry anyway, and we never did I don't think. Of course my grandmother was a great cook."

Robert Wimble remembered, "Will Kingsbury who was, very generous with his time, and very understanding—he was their blacksmith. At that time all the farmers had horses, and the lumber company had horses and they moved all lumber around their yard with horses. I can remember as a youngster, before I would get up in the morning, I'd hear the horses go down through cloppity, clop, clop, with their horses. Especially in the wintertime, you could hear the squeaking of the steel. They didn't have rubber tires, they had steel wheels, and they had a lot of these Army wagons. I think they were left over from WWI; they used them for the lumber, to move the lumber around. This was before trucks became as prevalent as they are now. We might have at the most twenty cars going up and down route 100 in those days.

"It was interesting for me to, to see the transition from the blacksmith where everything was done with horses and he would make a lot of wagon wheels and repair them, and he would make things out of metal. He had a forge there. I used to turn the crank and to make the air to go through to burn the coal and he'd hammer just like an old fashioned smithy, you know. They even had a place there where they used to shoe oxen."

Mills and Water Power

At the upper end of Moretown Village, at the bridge on Route 100B, the Mad River's elevation is 600 feet above sea level. At the lower end of the village, its elevation is 560 feet. That 40-foot drop, combined with two narrow gorges, and the collected water of most of the Mad River's watershed, account for Ward Lumber Company's long connection to the river. Water power was vital to the business from the day Hiram Ward opened shop in Moretown until the mills moved to Waterbury in 1962.

A typical Vermont lumber mill of the late 19th and early 20th centuries had a timber crib dam on a stream or river. The Wards had two. Using Moretown's geography to their advantage, they had a dam at each of two narrow cascades in the village. One dam served the Ward Upper Mill; the other served the Ward Lower Mill. Situated less than a half-mile apart, the dams held back the river in two steps. With this arrangement, the company benefitted twice from the same water. This was particularly helpful during dry summer months, when the river's volume was minimal. The water held in the upstream millpond ran through the penstock and turned the turbine for the Upper Mill. After passing through the turbine, it continued downstream to the second millpond. Once that pond filled, the water could flow through the lower penstock and run the machines of the Lower Mill.

Both Ward dams were of traditional New England timber crib construction. A timber crib dam was exactly that—a box made of timbers, usually hemlock or spruce logs. The structure was wedge shaped, its vertical face aiming downstream. Usually the crib was filled with rocks for stability. The upstream side of the dam was faced with boards. This deck of planks was essential to smooth operations. As the dam stopped the river's natural flow, the millpond filled behind it. Thanks to the planks, excess water smoothly flowed over it. In summer, the dams had flash boards installed on top. With these removable boards, the water level could be raised even higher than usual.

On the east side of the river were the mills and their equipment. At each dam, water from the millpond dropped down through a penstock. The penstock was much like a sloping wooden silo that directed the water to a small chamber holding the waterwheel. Unlike the wooden structures of quaint country pictures, these turbines were made of steel. Each turbine had a drum at its center point with attached metal

The Upper Mill with the old covered bridge behind in the background

The old Upper Mill as it was with the timber crib dam circa 1930s.

Courtesy of the Moretown Historical Society

blades fanning out from it. The water, with the force of its drop from the higher elevation pressed against and turned the blades, rotating the turbine. A heavy steel shaft extended from the center of the turbine up to the mill building. The rotational energy of that shaft, driven by the flowing water, ran the sharp saws. Belts, bearings, more shafts, and gears—a vast system of them—transferred the energy from the shaft to the machines.

This straightforward system served the Wards for decades. The Upper Mill had a horizontal steel turbine that stayed still in place long after the mill was gone. It even survived the 2011 flood. The Lower Mill had both horizontal and vertical wheels. The vertical turbine rested on pieces of lignum vitae—very hard wood with a natural oil in it. On the lignum vitae the turbine could spin freely.

This water-powered system required maintenance and some unusual repairs. Rocks and gravel generally settled out in the Upper Mill pond. Branches and other debris floated, if swept into the turbine, they could do extensive damage to it. A rack installed at each penstock's upper end caught most of the material. Raking the branches and other objects from the rack was part of the mill routine.

From time to time the penstock needed refurbishing. Owen Ward remembers the unusual and dangerous procedure for its repair. "This was a six-foot diameter penstock as I remember. Every now and then it would rot out and we'd have to make another one. We'd dress the hemlock. We had to take off the hoops that went around it." Like silo hoops, these tightened to hold the wooden frame together. The replacement boards were thick wedge-shaped pieces and had to be precisely cut.

Once the pieces were prepared, the frightening work began. Up at the surface, planks were placed across the penstock's opening, those boards kept water from rushing into it.

Owen explained, "You'd work inside the pen-

As the upper mill site looked in the fall of 2011 where some of the foundation posts are visible, as well as the outlet from the turbines. Notice how the river has eroded the bank on the left.

stock. As you worked down towards the waterwheel it was always hairy. You couldn't help but remember that we used boards to shut off the water." If the boards failed, a person inside would be trapped. "If you're working down underneath there right in the waterwheel itself, you're wondering. You know that if the water starts coming in you've had it, you have no way out of there. That's one of the things that never did happen—but everything else that could happen did happen."

As a boy of 12 or 13 years old, Owen assisted with an unusual repair. The planks on the upper side of the dam were essential to the dam's proper function. Rocks, branches, logs, often battered against them. A hole in the dam would get larger and lead to bigger problems. One summer in the 1930s, Kenneth Ward was concerned about some apparent damage to the planking—several feet underwater. On a warm day, Owen put on his swimming trunks and waded into the mill pond. Several sand bags were filled—bags that doubled as divers' weights and repair materials. As the adults directed him, Owen would get in position above a suspected hole, then, holding a sandbag, he slipped underwater, feeling the planks with his free hand and feet. When he found a hole, he would stuff the sand bag in it. With repeated descents, the leaks were plugged.

Early springtime was particularly perilous at the mills with the spring ice flow. Holly Ward vividly remembers a scene from his childhood, watching a remarkable effort to protect the town from flooding and the dams from damage. In the spring, then as now, days got longer and warmer and winter's ice started to break up on the river. Flowing downstream, great chunks of ice collided with the dams.

"The ice could be twenty inches thick. It would get

Holly Ward in the clapboard mill with Merrill Reagan's augur

stuck at the dams," Holly recalled. There was considerable risk that these huge pieces of ice would tear the planks of the dam. As the ice piled higher and higher, it acted like a dam itself; the water it held back could flood the village. "At the Lower Mill the ice would built up terribly with huge, huge cakes of ice.

"Merrill Reagan was very important at the mills. I remember watching him in the springtime when I was a little kid," said Holly. "He would go out and walk across the ice, on top, ice that was ready to topple over the dam. He would walk out there with dynamite and an augur and a long fuse. With his big augur he would cut a two- or three-inch diameter hole maybe eight inches into the ice. The ice was moving, mind you, waiting for something to happen. He'd put the dynamite in and follow the fuse back." Back on shore, the fuse was ignited. When everything went right, the dynamite exploded and the ice jam was blasted to bits. But, Holly remembers, "A couple of times the fuse got in the water and wouldn't light and he'd go back out there.... It was astounding!"

The Wards managed the dams carefully in order to run the business in summer months when the Mad River's flow was minimal. Temporary extensions, called flash boards, were installed at the top of each dam to raise its height. In summer, the ponds took longer to fill, but with the flash boards the ponds could hold greater volumes of water, allowing some extra time for the mills to run. Using the available water power required some flexibility in the work schedule.

Holly explained, "Before we had auxiliary power, in the summer when the water was low, they would go to work at 3:00 in the morning. They would open the gate and run to 7:00 or 8:00, using up the water. Then they'd go home and then wait for the pond to fill up again. They would come back at 3:00 or 4:00 in the afternoon and work as long as they could then."

To increase the reliability of production, Ward Lumber Company added auxiliary power: a steam engine in the Lower Mill first, then a diesel engine in the Upper Mill, and later, electricity in both. The smokestack of the wood-burning steam engine in the Lower Mill is evident in photographs from the 1930s. The steam engine helped, but bitter winter weather still posed challenges, even in the mills' later years in Moretown.

"In the wintertime, it'd be all frost and sometimes you couldn't get the mill started," recalled Owen. Frigid temperatures congealed bearings that needed to move freely. Without them, the belts spun uselessly on the pulleys. "We'd get warm ashes and put them onto the belts. We'd be there trying to feed warm ashes onto these belts at 4:00, 5:00, 6:00 in the morning trying to get the mill started by 7:00. Sometimes you couldn't get the mill started until 10:00 in the morning." One of Owen's initiatives at Ward Lumber was adding electric heat to the Lower

Scattered chunks of the upper mill foundation lie where time and floods have deposited them. The remains of the steel waterwheel at the upper mill site are still in place. If you look behind the tree, you can see the shaft of the waterwheel with a wheel on top (upper arrow). Part of another wheel is still there at the lower arrow.

Mill—an addition that relieved this problem.

The steam engine also had occasional winter problems. "The boiler room was right down by the flume. Many mornings they'd call me up and say, 'We can't get water into the boiler because the pumps are frozen up.' The injector wouldn't work—the injector injected water into the boiler. As I remember, you had to have 15 to 20 pounds of steam. But if you couldn't get water in the boiler...." In this predicament, Owen had to decide whether or not to build a fire. "If you don't have a fire, you can't get water in the boiler. On the other hand, if you build a fire, and you can't get water in the boiler, you're going to blow the boiler up. It got hairy lots of mornings. So, we would call three or four guys and have them come down. Then, three, four, or five of us would stand there and bail buckets of water—it was sometimes 20 or 30 degrees below zero. We would pull the buckets of water out of the foam, take them and pass them to another guy, he then passes them to another guy. He would put them in where the water was supposed to go." They would keep adding water, trying to, "keep it going until you could get your lines thawed out so things could get back to normal without hopefully blowing the thing up."

"When you got through, you'd be nothing but ice—after pulling that water out with ropes, taking a rope and throwing it in, having a rope on the pail, pulling the pail up, giving it to another guy. You couldn't worry about getting wet—I mean you're doing hundreds of pails of water and it would be −30 degrees out. The guys inside wouldn't do so bad, but the guys outside would be nothing but a sheet of ice. That severe weather condition was when you had that kind of problem because if it wasn't that severe, things wouldn't freeze like that."

Water power was important to Ward Lumber Company operations in Moretown even after auxiliary power sources were installed. The dams were still in place when operations moved to Waterbury. Gradually, through the 1970s, the dams gave way. Remains of the waterworks though are still evident a half century later.

The lower mill's dam was located where the rocks narrow and the water falls. The right side of the river, where the water is calm, leads to the penstock.

Remains of the lower mill exposed after TS Irene in 2011

The US Geological Survey maintains water-flow monitoring stations along Vermont rivers. The Moretown gauge is just downstream of the village.

The average low-flow months on the Mad River are August and September. During this time, flows typically range from 40 to 75 cubic feet per second.

In the 1938 flood, the flow reached 18,400 cfs. During the August, 2011 Tropical Storm Irene, the flow was recorded at 23,600 cfs.

Belts, Babbitts, and Bearings

Below the floor where lumber was cut, were the belts, shafts, and bearings that carried the power from the turbine to the machines. These interconnected parts needed to run smoothly for the saws and other machines to do their work

The shaft from the turbine was the centerpiece of the system. From it, other turning shafts, belts and pulleys carried the energy to the saws. Sets of belts could be engaged or disengaged to direct the energy to different machines. Early belts in the mills were made of leather—later they were rubberized. Some belts were over a foot wide and ran over a distance of 30 feet—so they were actually 60 or 70 feet long. At the end of each shaft was a bearing allowing motion between two different parts. Minimizing friction was essential to the bearings' efficiency. They needed to be smooth-running and lubricated so that they would not seize up when spinning metal on metal produced high temperatures.

Belts, wheels, pulleys

Back in 1839, Isaac Babbitt of Taunton, Massachusetts, devised a bearing design that was a boon to mill operations. It was such a success, that "babbitt bearings" and "babbitting" continue in mill vocabulary. Instead of the wood bearings of earlier times, Babbitt's design used an alloy of lead and tin where the parts met.

"The horizontal wheels were suspended on babbitt bearings that required lubrication via wicks or rings that revolved when the shaft turned, picking up oil from inside the housing," explains Owen Ward. "As the oil surfaced, it flowed through the length of the babbitt box, due to channels cut in the babbitt, thereby lubricating the load area and avoiding friction that would cause the box to heat up. The box, was commonly called a hot box. This was a common way of supporting shafts with pulleys and/or shivs and belts that powered some types of machinery."

Replacing the babbitt needed to be done regularly. "Now the difficult part is, it takes a real millwright in order to do babbitting on a bearing. It's very difficult to do because you've got to take out all the old babbitt, and you've got to know how to get it to flow to the right amount of babbitt on the cap and on the bottom. You use a chisel to smooth the babbitt all down. Then you put it in and you run it for awhile and get the oil on it—then you find there may be a high spot which is a little higher than the rest. You have to take it back out again and smooth it again where it's high and then put it back in. It's very difficult."

At the Mills

For most of Ward Lumber's 70-plus years in Moretown, the company had mills at both ends of the village—the Upper Mill and the Lower Mill. The processes at them were similar, although each one specialized and had equipment for its particular products.

Through this long history, both mills saw changes. Sometimes these changes were prompted by disasters. Fires destroyed the buildings and equipment at the Lower Mill in 1935, and the Upper Mill in 1955. Some changes were driven by the Ward's business decisions to develop products for specific markets, such as the burgeoning furniture market.

From the late 1930s, the Lower Mill mostly cut hardwood. The Upper Mill historically dealt with softwood, but in later years, it cut parts for hardwood furniture including chair seats and crib rails. A bolt mill was also installed at the upstream site.

The Lower Mill was producing around 3,000,000 feet of lumber each year in the 1950s, the Upper Mill was producing 1,500,000 to 2,000,000 feet at that time. In those years, between 40 and 50 men were usually employed between the two mills—moving logs, running saws, maintaining equipment, sorting, stacking, and grading boards.

The first step at the mills was the same—dealing with arriving logs. Both mills stood by the river, with the main road passing above them. The arrangement for delivery of logs used gravity as well as muscle power. Logs were rolled off the wagons—and later off the trucks—into the yard. Having an adequate supply of logs to meet orders was essential. Sometimes, the log piles would be nearly 20 feet tall, reaching almost all the way up to Route 100B.

A winter morning at the lower mill, the hot pond is on the right front of the building with steam rising from it. On the left you can see a pipe that carried wood chips from a wood hog under the mill. A pipe called a separator separates the chips from the blowing air. On the left, coming down the hill, is a line of logs awaiting their bath.

The old lower mill in the winter—before it burned in 1935

The rebuilt Lower Mill in the winter of 1938 with steam coming out of the smokestack

Lower Mill

The Lower Mill, at the downstream end of the village, stood at the site of Hiram Ward's first Moretown venture. An earlier mill on this site burned in May 1887. After the fire, Hiram purchased the site, built a new mill, and by 1889 was proprietor of a "box factory and grist mill" there. From this start, the Lower Mill became one of the anchors of the Ward business.

Recognized as one of the finest in the region and equipped with modern machinery, the Lower Mill was productive and an important center of employment for years, even during the Depression. Its prominence in the local economy made its loss to fire in 1935 especially devastating. The fire, set by an arsonist, totally destroyed the mill building and equipment. Diligence of the attending fire departments was credited with saving much of the lumber in the yards. The fire was a terrible blow to the company, but the Wards rebuilt. Men were soon back at work producing lumber.

The new mill had a new wood burning steam engine to supplement power from the river. (In later years, electricity also powered the mill's machines.) The steam system offered an extra benefit—a steam pipe from the boiler heated an outdoor hot pond. Contrary to its name, Owen Ward remembered the pond as closer to tepid than hot. Before logs went into the mill to be sawn, they were dunked in the pond to soak for a few hours. The water loosened the bark and cleaned out embedded dirt and rocks, picked up when the logs were skidded. After their soaking, logs were pulled out by a hook on a big chain connected to a bull wheel. They were placed on a deck that was level with the carriage, in position to move to the saw.

The centerpiece of the rebuilt Lower Mill was its band saw, a powerful tool that could cut even very large diameter logs. The band saw's blade was a long metal band with sharp metal teeth along one edge. It ran around two big six-foot diameter wheels—wheels that needed to be positioned precisely for the band to have the correct tension for efficient operation. The teeth in the band were shaped to rip through the wood and discharge the sawdust produced by each bite. The band saw cut a relatively narrow kerf through the wood. With a thin kerf, less wood was lost as sawdust.

The band saw, for all its advantages, needed considerable maintenance. Sharpening its teeth was a constant part of the mill routine. The bands for the Lower Mill's saw were 36' long and 11" wide. To keep up with the work load, the mill owned several bands. As one dulled, which was nearly daily, it was carried up through the ceiling on pulleys to the upper level—the filing room. The filers pulled it off the pulley, stretched it out, and ran it through a grinder.

In the meantime, they had lowered another band and the saw was again whipping round and round on the wheels below them. Between the handling, sharpening, and reinstalling, the maintenance of the saw was time consuming.

In the heavy business of the lumber industry, logs must be moved efficiently. The logs, not the saws, move through the mill. They pass through the saw riding on the carriage. In early saw mills, the carriage was a wooden frame, resting on wheels. The sawyer rolled the log onto the carriage and fastened it down. The carriage was set in motion with ratchet works connected to the saw and waterwheel. The carriage moved the log through the saw, then it shifted to make its next pass and allow the saw's next cut. While still performing their critical carrying function, later carriages were stronger, faster, and more automated than the early devices.

In later years, the Lower Mill's new carriage was driven by a steam piston. The carriage's deck sat atop wheels, trucks, that rolled on steel rails. Levered dogs with metal teeth, held each log in place. Once loaded on the carriage, the log moved through the saw several times, each time the teeth of the band whipped through the wood, ripping off the desired dimension of lumber.

In the 1950s, one of Owen's initiatives was the purchase of a new machine. This steam-powered lever arm turned the logs on the carriage. With it, a log could be rotated to enable a range of cuts. After the first cut removed a slab from the log, as the carriage pulled the log back, the sawyer looked at the face of the log. He judged the highest quality dimension that could be cut from it. This steam lever allowed him to easily flip the log, so it could be cut to its greatest potential. "It was helpful for every single log," Owen noted.

The steam lever increased efficiency. "It was just fantastic," recalled Owen. Hardwood is heavy. "Thirteen pounds to a foot is what a hardwood log weighed," he explained. A log calculated at 200 board feet by Vermont Rule weighed about 2,600 pounds. "It would take that log, pick it up after you took off one side, turn it over and lay it down flat." Heavy timbers for railroad ties were a snap with the steam lever. "We used to saw the hearts for ties for the railroads. We'd make ties that were like 6" × 8" × 8 feet.' We also made them 7" × 9" × 8.5 feet. Now those ties were all green. They rolled them out, they were heavy. You're talking about something that weighs almost 300 pounds."

Sid Atkins (with mustache) has worked for H.O. Ward and the Ward Lumber Co. for 54 years. He has been thrifty and dependable. His helper at the clapboard mill is Minerd Fielber is 73 also and has worked many years for the lumber co.

—Merlin Ward, from the Family Album

Running these machines required skill. The men who worked in the mills honed their abilities for certain tasks. One job of the time, involved "riding the carriage." Owen explained, "It was a shotgun feed carriage, better than the hydraulic systems as far as I was concerned. A shotgun feed had a huge pew that was filled with steam. There was a big piston and it was hooked to the carriage and steam controlled. The sawyer would move a lever and it would open up a valve. That would determine how much steam you let into this piston. The more steam you let in, the faster it would go and the faster it would feed down through. Then it'd come back and if you were outside you'd hear it go 'Bing!' You'd hear the pop of the steam stopping."

One day, it didn't stop. "It shot him and the whole carriage along with him across the room. It tore out half the mill down here, right straight through," Owen remembered. "The guy that rode

A view of Moretown and the lower mill from across the river. The road behind the logs is Main Street, where logs are unloaded from trucks, then stacked awaiting their scaling, sorting, and sawing. What may not be evident is the depth of the drop from the road to the mill allowing gravity to help with the unloading of logs as well as the logs progression through the process into the mill.

it was hurt. It also tore out all of that equipment. The mill was shut down for about a week. We put it back together again and they were trying it out. It did it again; it went right down through again." The cause was finally determined, a bolt had dropped into the valve and the valve wouldn't shut. "The whole mill was shut down again."

When the mill shut down, it was a hardship for everyone. "We tried to keep the guys working all the time, all those people had to eat. But it was tough for the company to make those hours," recalled Owen.

While the band saw did the work of long cuts through the logs, there were further steps to produce the finished board. After running through the saw, the boards still had bark on their uneven outer edges. The next machine, the edger, cut off those edges, giving the board its rectangular profile. The trimmer next cut off the ends. Like the sawyers, the men running the edger and trimmer used their judgment to get the maximum grade from each board.

After travelling through these machines, the edged and trimmed board was dropped on a bench, where the grader, with his trained eye, looked at each board's dimensions and judged its grade. The boards were then sorted according to grade. At first, this was done manually. Later, an automatic edge sorter, an impressive system that ran nearly 400-feet long, took some of the heavy labor out of this step. A chain on rollers transported boards from the mill, after they had been edged and trimmed, out to the covered storage area where it delivered them to appropriate piles based on their grade and thickness. Once in these piles, the lumber was stacked with stickers; the stacks were then moved by forklift for longer term storage.

At the Lower Mill, during later years, the lumber then generally went in one of two directions. Some hardwood was stacked in the mill yard right outside. From there it was shipped out to fill orders from furniture companies. A few companies, including Vermont-based Ethan Allen, owned kilns for drying the hardwood they purchased. Much of the lumber was taken to the storage yard to dry. Ward Lumber owned a large flat field a short way up the Moretown Common Road. Hundreds of thousands of feet of boards were regularly drying there in tidy 16-foot stacks. Drying usually required several months. Once the boards were dry, many of them went to the Upper Mill to be cut for furniture parts.

A Piece of History

With the band saw, the lower mill could cut very large diameter logs. The Wards had some very old massive hemlocks on a parcel of land in Warren. The St. Lawrence Seaway was under construction at the time, in the late 1950s. These huge hemlocks fit the contractors' requirements for timbers to provide solid bases for their cranes. The hemlocks were cut at the Lower Mill and shipped off to Canada.

The lower mill, before the additions and extensions were added

The lower mill, before fire destroyed it, circa early 1930s.

I liked him. I think they [the Wards] liked me. Merlin come in one time, I was working in the mill on the clipper they call it. You drop the board on it, cut, squared up both ends, it went up over automatic. I had to take the things off the rolls after it sawed from the log. Had a hog behind me and all the slabs went down through that… [Merlin] got so he knew me. So he stayed away from me for part of a day, [after they had a row] and let me cool down. He come along, slapped me on the back one day, "Gee whiz," he says, "You got an awful temper." He says, "I like you just the same." And that's the way I felt about them.
— Guy Livingston

…B.S. Ward… he come down and got me one day to go up to the cemetery to level the cemetery off, the grade there. I went up and helped him. On the way home, he said, "Boy you done awful well. From now on," he said, "I'm gonna give you twelve cents an hour."

I ran all the way home to tell my folks that I got a two cents an hour raise. Yup. I thought that was a big treat, two cents.
— Warren White, around 1940

The lower mill as photographed by GHW on May 6, 1931

Upper Mills

The picturesque Ward Upper Mill stood just downstream of the bridge at the southern end of Moretown village. In the early years, the Upper Mill was a softwood mill. Unlike its downstream cousin, this mill had a circular saw—a big 52" circular saw. Circular saws were designed earlier than band saws and their straightforward operation had considerable merit. The teeth on a circular saw could be replaced, they were easier to file, and the saw was a stable piece of equipment. As the Lower Mill supplemented water power with the steam engine, the Upper Mill added a diesel engine to provide more consistently available power.

The original upper mill along the river with its crew

At this mill, millions of feet of dimension lumber were cut for construction projects such as homes and businesses. In the 1950s, the Wards had contracts to provide lumber for complete houses. The lumber for each house was loaded on a truck and taken to the building site, usually in Burlington.

In later years, the Upper Mill continued to saw softwood and it also produced pre-cut furniture parts. Seasoned hardwood boards that had been cut at the Lower Mill were brought here and cut to the specifications of furniture makers.

In the 1950s, the Wards added another mill, on the upper side of Route 100B. Along with increasing productivity, this mill's site had the added advantage of providing more storage space for logs. At the original Upper Mill, the distance between the building and road was quite small, making storage of sufficient logs a longtime challenge. The new mill had ample space. At first, this upper Upper Mill was a softwood operation.

The addition of the bolt mill at the Upper Mill in the 1950s was an effective and successful innovation. Although it sounds like a paradox—the bolt mill could cut high quality lumber from lower quality logs. Top grade logs are long and straight with few knots. Maple, beech, and other hardwood trees of the northern forest do not always grow tall and straight. Even with a twist or bend or branch, many of these still have tremendous potential if their clear lumber is cut from between the flaws. The bolt mill made it possible to get that higher grade lumber.

At the bolt mill, full length logs were carried on a conveyor under an automated chain saw. The chain saw cut them into short lengths 40" to 6' long. Next, these bolts moved to the bolter saw that cut them into boards. From there, they went through several rip saws that cut them lengthwise. The saws could be set to produce pieces of specific dimensions.

Chair legs, for example, were 2" × 2" by 20" long. These pre-cut pieces were shipped to high-

Walter Reagan, foreman at upper mill. Now learning how to operate the new 125 horsepower diesel engine installed in 1939. Shortage of water…
—Merlin Ward, from the Family Album

The new, higher, upper mill built in the meadow across the road from the river. The photo is dated "printed the week of June 9, 1956."

end furniture factories. At the factories they were shaped with lathes and saws, sanded, finished, and assembled into furnishings. "We made thousands and thousands of chair legs," Holly Ward recalls. Companies used Ward Lumber products in their fine furniture.

Fire has always been a problem at lumber mills. Once one starts, with the abundant fuel on hand, it can quickly spread. The Lower Mill had its great conflagration in 1935. Fires also took their toll on the Upper Mill.

On July 26, 1955, a fire started in the Upper Mill in the early morning hours. Fire departments from Waitsfield, Warren, and Montpelier came to the aid of the Moretown crew. As the Montpelier newspaper reported, "Lack of water hampered the work of the firemen. Water was pumped from the Mad River which is low at this time of year." The cause of the fire was not determined. Within hours, the historic mill and its equipment, including the furniture production machines, were gone.

After the fire, the Wards constructed a new building at the site of the historic Upper Mill. A Porter Rough Mill System was installed here. This system used lower grade hardwood boards, boards from 6' to 16' long, and cut out the defects. Its final products were clear, cut-to-size furniture parts such as backs and seats for chairs, rockers for rocking chairs, and table tops.

On a bitterly cold February morning in 1961, fire struck again. This new upper mill was barely five years old. The fire is believed to have been started by the wood-burning furnace. The furnace was functioning normally when the night watchman checked it at 2:30 a.m., but a little later the building was consumed in flames. A 700-gallon oil tank exploded, making the situation even worse. Electricity was cut off throughout the village. There was no loss of life, but the fire dealt another blow to the company.

DEVASTATION AT MORETOWN—Friday after $50,000 fire destroyed Ward Lumber Co. Flames broke out at 3:30 a.m., Resulting in power failure that closed village school for the day. Firemen were hindered by freezing temperatures.

—Burlington Free Press, Feb 2, 1961

```
          1938
     RETAIL PRICE LIST

#1 Spruce Dimension D4S up to 16 Ft.----$33
#2    "         "     , where possible
                           rough--------$20
                           D4S----------$24
Matched Spruce
                           1 X 4--------$28
                           1 X 5--------$30
                           1 X 6--------$33
#2 Matched Spruce          ------------$22
#3 Spruce & Hemlock beards, P1S & C-----$20

#1 Hemlock Dimension  D4S up to 16'-----$30
#2    "         "     , where possible
                           rough--------$20
                           D4S----------$22

                CLAPBOARDS
6" Extras---------$45    5" Extras -------$30
6" Clears---------$40    5" Clears -------$25
6" 2nd Clears-----$30    5" 2nd Clears ---$22
6" Cottage--------$22    5" Cottage ------$15
6" #2-------------$10    5" #2 -----------$8
5½" Extras -------$40    4½" Cls & Extras -$25
5½" Clears -------$35    4½" 2nd Clears --$20
5½" 2nd Clears ---$29    4½" Cottage -----$14
5½" Cottage ------$20    4½" #2 ----------$8
5½" #2 -----------$10
                4" Cls & Extras --- $25
                4" 2nd Clears----- $20
                4" Cottage-------- $14
                4" #2------------- $8

Red Cedar Shingle---#1-5X------$6
                ---#2-5X------$5
                ---#3-5X------$4

Spruce Finish--4½¢ & 6½¢ per ft.
Hemlock Finish -- 4½¢ per ft.
Hard Pine Flooring -- $60
1¼" Spruce Porch Flooring -- $40
#3 rough boards ---- $16
```

This price list will give you an idea of the variety of lumber from the mill.

ACCIDENTS

With sharp saws, heavy logs, and a labyrinth of moving belts and shafts, mills were dangerous places. Accidents occasionally happened, resulting in injuries or even death. Tragedies deeply affected the close knit community.

Roger Hayes was just 22 years old in 1958 when he was crushed by a heavy maple log at the Lower Mill. Hayes had his own truck and was delivering logs. On that February day, he was prying an especially large maple log loose from the load on his truck. As he worked, he apparently slipped. Owen recalls, "he fell down, and that log, that big log, came down." Owen and others rushed to his aid. Roger Hayes' brother, Richard Hayes, was operating the forklift, he got there with the machine, but it was too late. "It was tragic; a young guy like that just started out on his own."

Accidents occurred inside the mills, too. Cy Kingsbury, a longtime Ward foreman, he knew the equipment and routines. In a freak accident, he was killed while working with the bandsaw. "It was so sad and just unexplainable," Owen remembers.

Owen also had a close call in 1952. He was grading lumber, standing on one of the tall piles of boards. "It didn't have anything on either side of it, just happened the other piles were gone and I was on top of this pile way up above the track. Somehow or other I caught my heel." He fell to the ground, and was knocked unconscious. Employees blew the steam whistle—the signal during an emergency. Help arrived and Owen was transported to the hospital by ambulance. After time in the hospital and convalescence, he was back to the mills.

GRADING OF LUMBER

The National Hardwood Lumber Association (NHLA) established *grading standards* for lumber over a century ago as *grade* (the quality of the wood) greatly influences the price of lumber. Using these standards, both the buyer and seller have a shared understanding of lumber quality. The basic unit in the industry is a *board foot*, a volume of wood one foot long, one foot wide, and one inch thick.

In grading lumber, the grader calculates the amount of clear wood in the material being evaluated. A board that yields larger, longer pieces of clear wood is of higher grade and value than one that has smaller clear areas.

All logs are graded before entering the sawmill to determine what they will become—bolts, lumber for construction, furniture, and so on.

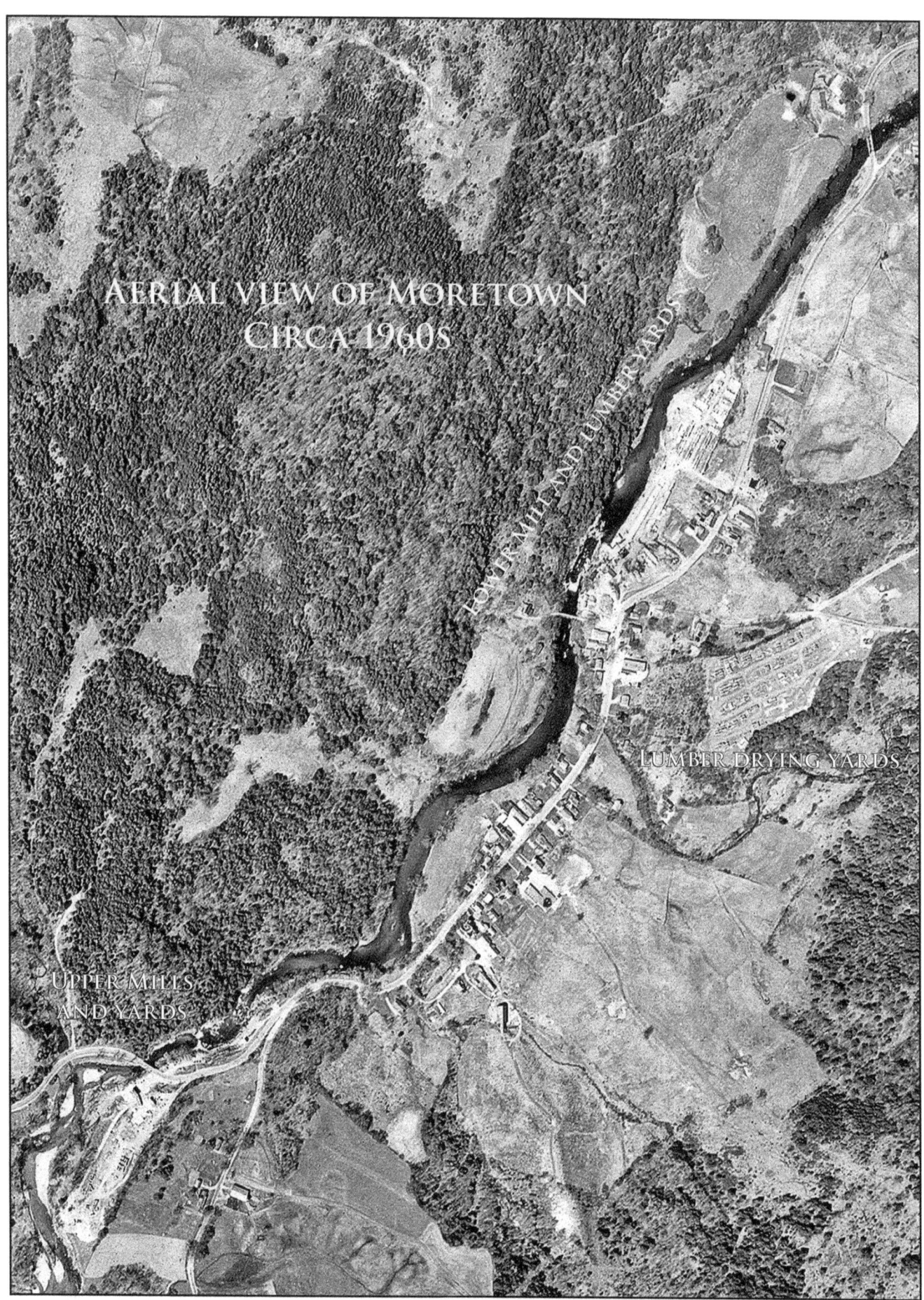

Owen, Merlin, and Holly in the office, April 1959

Monday Morning at Mill #3

It's 6:55, and the whistle just blew,
The boss is busy checking the crew,
There's Bo-bo, there's Farnsworth,
There's Carlos and Don,
Brooks and his crew have already gone.
Here's George and here's Irish,
And Freddie and Frank;
Livingston's coming, and Norm's by the Tank.
Pat's gone home—but he'll be back.
Peanut is firing 'cause there's smoke in the stack.
The steam is up but the water is low,
Hot pond's dry, and the Lift won't go;
Eugene's out—hasn't got back,
And Merlin's about ready to blow his stack.
Lumber to go—don't know how,
Cause the Dozer's gone up to the mountain to plow.
Then Holly blows in and says with a frown,
I can't run—the Bolter's broke down.
Mill should start, but don't know when,
Perley's gone; can't find Ken.
Filere can't file, cause his rig is broke,
And that for certain is no big joke.
Crane is stuck, and the hog is plugged;
Over the weekend, some of the boys got jugged.
Edger is dull and George can't dog,
The whole damned yard is like a bog.
Owen takes a look and says, "Oh, well,
    I'm heading to camp...
    Let it all go to hell!"

A poem, written by a local wit, was framed and hung on the wall of the Ward Lumber Company office for many years.

Surveys and Changing Land Use

"For log jobs, we had to know exactly what land we owned," explained Owen Ward. The extensive landholdings of Ward Lumber were considerable assets to the company. The land provided a steady supply of logs for the mills. By managing the cutting, the land had the potential to continue producing timber into the future. The selective cutting of larger trees had a double benefit—sending those to the mills, and also encouraging growth of the smaller, younger trees.

Sending loggers onto a piece of property to cut trees requires a clear understanding of the location and boundaries of the parcel of land. Cutting logs from land you don't own carries significant penalties in Vermont. Trees have value—property owners are entitled to compensation and punitive damages if their trees are cut without permission, even if the cutting is accidental.

The over-30,000 acres owned by the Wards covered more area than all of Moretown's 25,743 acres. Unlike Moretown, though, the Wards' territory was not one large parcel with a few boundary lines. Instead, it was comprised of scores of individual pieces—some 30 or 40 acres each, others of hundreds of acres. Some of these had property lines that were set out in original town charters—Moretown's charter dated back to 1763. Other property lines were described at the time of past sales. As Hiram Ward had acquired a lot of land in the late 1800s, many of the deed descriptions dated from that era.

"It took a lot of detective work and a very intelligent surveyor—Paul Bigelow," remembered Owen. Through the years, Owen and Paul Bigelow tracked down miles of property lines. Their quest led them along old stone walls, old fence lines, and along the banks of the Mad River. "Sometimes a deed would say 'along the thread of the river,' but the river had moved. I remember going to a property in Warren below the falls on Route 100, the deed said the corner was at a 'bushy topped maple,' but the deed was written 50 years before. Paul and I searched and searched, and I think we finally found it."

Ward forester, Leo Laferriere, remembers working with Paul Bigelow's maps in 1960s, and the way the surveying could be done with aerial photographs purchased from the U.S. government.

"Paul put all the boundaries on the aerial photographs. Now they had pictures of all the lands, and the boundaries. I found out after using these maps over four or five years that they were quite accurate. Paul had to do a certain amount of photo-interpreting, say, locating a road (in the photo and on the map), ridgelines, and so forth.

"The way you use these maps is to overlap them, and you use a stereoscope and it becomes three-dimensional. The dark areas are softwoods: balsam, fir and the like; the light areas are deciduous trees: sugar maple, white ash, beech, things like that. So it's easy to type where the changes are obvious. Like in an area where the trees are all similar in terms of species, composition. And I can tell what they are without going out there, say here's a high elevation so it would be spruce, balsam fir, maybe a little hemlock.

"After you've used a huge stereoscope for a while you can pick out subtle differences, for instance in one area there might be large trees, in another small trees. I could take this photo, and go to this plot of land, walk into one section and identify two trees (from the map). I've been able to do that. This is how useful these photos can be. You can pick out the plantations, because they're all in rows."

Beyond the "bushy topped maples" and other features, land use in the region had changed a great deal since the Wards acquired most of this acreage. In the early years, outside of the village centers, rural Vermont land was either in agriculture or forestry. After World War II, that began to change.

Mad River Glen, opened for skiing in 1948, was one of the first ski areas in Vermont and was built partially on land sold by the Wards for its development. In 1958, Sugarbush Valley, opened in Warren. In 1962, Walton Elliott purchased Ward acreage and announced development plans for Glen Ellen ski area. With these new winter attractions, and improved access throughout Vermont with the arrival of the Interstate highways, the state's population grew. In the Mad River Valley, the tourist economy also grew as restaurants, inns, and other services opened. Remote land became more accessible with improved town roads. Demand for land for subdivisions, new homes, and developments changed the long-standing rural landscape.

SURVEYING THE LAND

Establishing the property lines on the ground by marking them clearly was an important part of surveying Ward land.

Owen remembered some of the sleuth work involved. Boundaries are often marked with blazes cut into living trees. After a few decades, a blaze is not always easy to confirm. Following deed descriptions, Paul Bigelow and Owen often walked along old property lines, keeping alert for signs of old blazes. When they saw a likely candidate, Paul had a technique for confirming whether it was a blaze or a blemish. They used an axe to make a horizontal cut in the tree at the base of the mark. Then they made a vertical slash above and connecting to it. They could then see the darkened mark of the blaze (if it was one) and count back the exposed growth rings to confirm when it had been marked. If this matched the documentation, they knew they had the right line.

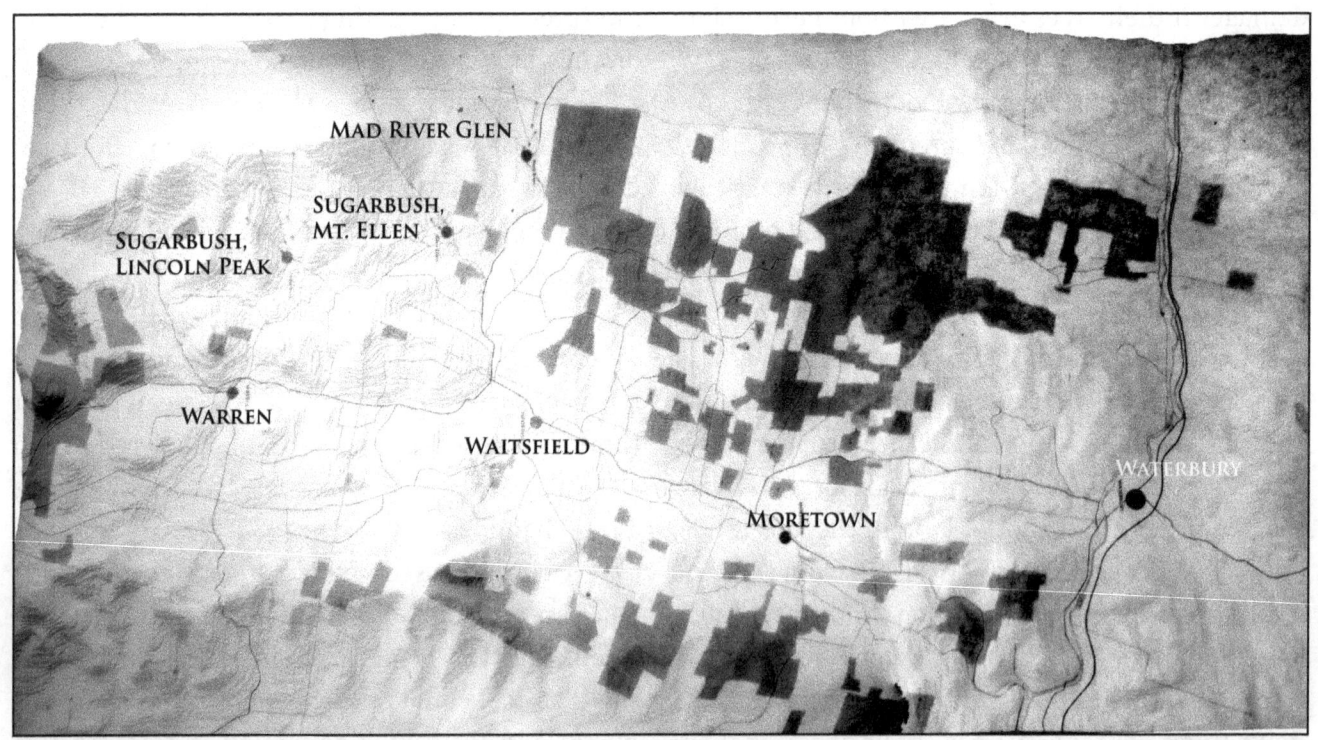

A relief map of land owned by the Ward Lumber Company in 1968—from north of Waterbury, south to Granville, east to Northfield.

Move to Waterbury

In 1962, one year after fire leveled the Upper Mill for a second time in a decade, Ward Lumber announced plans to relocate the mills to Waterbury. Many factors contributed to this decision.

Moretown's unique geography with its waterpower potential had drawn Hiram Ward to relocate from Dowsville almost a century earlier. The Mad River, even with its occasional fury, served the company well for many years. In the 1960s, though, there were other affordable, and reliable, options for power. A riverside location no longer had great benefits, but it still had perils, as floodwaters would certainly rise again.

With the company's substantial hardwood business, in the 1960s, the Moretown double-mill arrangement had become cumbersome. For the furniture contracts, logs were hauled through town to the Lower Mill and processed. Then the boards were trailered up to the drying yard on Moretown Common Road. Once dry, the boards were hauled back through town to the Upper Mill. Finally, the finished products were loaded on trucks and hauled through Moretown village once again as they went off down Route 100B to the train station or to the new Interstate-89 access in Middlesex. Moving the business offered the opportunity to consolidate operations—to coordinate all the mill activities on a single parcel of land. Lumber could then be shifted between buildings or sheds without the trek back and forth on town roads between the two ends of the village.

The new Interstate Highway system offered the Wards another good reason to consider relocating. When Hiram Ward's business took off in the late 1800s, railroads were the new transportation, opening a world of markets. In 1962, construction of Vermont's Interstate-89 was well underway. New segments were opening every year—trucks no longer needed to bump along secondary roads to carry Vermont products to other regions. Waterbury offered the double advantage of easy rail and highway access.

A Waterbury location offered yet another benefit to Ward Lumber—access to a larger timber-producing territory. A sufficient supply of logs was always critical in the lumber business. For decades,

Ward Lumber Company mills at the former Pilgrim Plywood location in Waterbury, looking south to Route 100 in the background. Today, this is the site of Green Mountain Coffee Roasters.

> **[NTPELIER], VT., FRIDAY, JULY 6, 1962**
>
> ## Ward Plans Revolutionary Lumber Mill In Waterbury
>
> WATERBURY — Owen Ward of Ward Lumber Co. said yesterday the firm plans to have a lumber plant here, revolutionary in the state, that will initially process four million feet of timber a year.
>
> The Moretown lumber company has taken up options on the Pilgrim Plywood plant in Waterbury, and has begun work preparing the plant for its own operation.
>
> Ward said Waterbury's new lumber plant will be "a bit of an experiment" in that its automated equipment will be set up in such a way that portions of the mill can be in operation while other portions are shut down.
>
> He said design of the mill will be so revolutionary that people will come from throughout the country to see it in operation.
>
> Plans are to have the plant in operation by November, Ward revealed.
>
> He said his firm "later on expects to employ more people from the adjacent area," indicating more job openings in Waterbury with the new plant in operation.
>
> "We definitely hope to expand here," he said.
>
> He revealed the firm plans to expend "considerable amounts" to renovate the old Pilgrim Plywood plant for its own use, and will make many exterior improvements on the building in addition to updating the interior facilities and moving in its own heavy machinery.
>
> The new lumber plant is one of three industries recently acquired by Waterbury, according to Craig Parker, president of the Waterbury Development Association.
>
> Due to extensive promotional work by members of the association and other interested townspeople, he said, two other empty industrial buildings in the town have been occupied.
>
> The town's only granite shed has been put into operation once again by Houle Brothers granite firm of Barre, and the former Cooley-Wright Manufacturing Company foundry has been purchased as a depot for the Valcour Chemical Company.

> ## Wards Buy Waterbury Plant
>
> WATERBURY — The Ward Company of Moretown announced the purchase of the Pilgrim Plywood Corporation plant in the Village of Waterbury.
>
> In the announcement, the president, Merlin B. Ward said "We feel that his plant is ideally located for our operations, since it is adjacent to the Throughway and on the railroad. The buildings themselves, with their large floor space, afford ample room for the company to combine the sawmill operation and furniture dimension plant under one roof, with plenty of room for dry storage."
>
> The Ward Lumber Company was able to acquire the large field adjacent to the mill, and officers of the company said they are thus assured of plenty of room for log and lumber storage and possible future expansion.
>
> The statement adds: "We have studied this move for a long time, and wish to thank the Waterbury Town and Village officers and the Waterbury Development Association for their cooperation in making this move possible. We are sure that this will be mutually beneficial." The Wards expect to have the Waterbury plant in operation in the fall.
>
> Officers of the Ward Lumber Company are Merlin B. Ward, president; Owen Ward, vice president and Holly B. Ward, clerk. The first of the Ward family to settle in Vermont, Hezakiah came to Waterbury in 1799, and settled on a farm property. In 1872, Hiram Ward began lumber operations in Duxbury. The father and grandfather of the present officers started the extensive reforestation project which has gained national fame for the operations of the company.

most of the logs for the company came from the Mad River watershed, and that area continued to produce much marketable timber. Waterbury, though, was in the Winooski watershed. Within a few miles of downtown Waterbury were tracts of land with new sources of hardwood, including the vast Nebraska basin toward Stowe. The Wards already owned productive forest land convenient to Waterbury.

In June, 1962, the Ward Lumber Company announced its purchase of the Pilgrim Plywood Corporation plant adjacent to the railroad tracks in Waterbury. (In later years, this became the site of Green Mountain Coffee Roasters.)

"We feel this plant is ideally located for our operations, since it is adjacent to the Throughway and on the railroad," Merlin Ward, President of Ward Lumber Company, said in a statement to a local newspaper. He continued, "The buildings themselves, with their large floor space, afford ample room for the company to combine the sawmill operation and furniture dimension plant under one roof, with plenty of room for dry storage."

That summer, Ward Lumber Company extensively renovated the Pilgrim Plywood building. It was a huge building, with close to 50,000 square feet of space. The saws, edgers, and other equipment from the Moretown mills were moved to the new facility. The Wards also invested in new machinery, bringing automation to their systems at the Waterbury plant. A new debarker improved efficiency at the starting end of the mill, and a kiln for drying the lumber improved the later process. The ample space made it possible to consolidate all Ward Lumber operations in a single building.

The Ward Lumber Company mill in Waterbury produced about the same amount of product as the Moretown mills—about 4,500,000 feet per year. This mill exclusively handled hardwood. The bandsaw and its associated machines were here, so were the bolter mill and pre-cut furniture production tools. In its expansive yard, forklift trucks moved logs and lumber. With the equipment and new layout, logs were sorted by specie and graded as they arrived. They could then be appropriately directed to either the bandsaw or bolter. This was a tremendous improvement in efficiency over the days when the mills and functions were at opposite ends of the village. The new location and efficiencies of consolidated operations served the company well.

Merlin's Speech

Presented at the Northeast Logger Congress
Woodstock, Vermont April 15, 1963
Panels "Integrated Product Utilization"

It was with considerable misgiving that I consented to appear on this Panel, for I am one of those people who shuns the limelight if possible. I have been afraid to face an audience ever since I was a small boy, and my grandmother made me speak "pieces" in Church.

But—they persuaded me, and here I am.

The Theme of this Congress is, "Integrated Product Utilization." These are stylish words in any man's language, and I had to ask just what I was to present in my paper. It is awkward to be on a Panel when you know that there are people in your audience who know more than you do, about your subject.

John Dork said that I was to give a brief history of our Company, showing the products we had produced, and how they are integrated at present, so, here goes:

Of course, in a History like this, personalities must appear, so please pardon so much mention of my relatives.

Well, our Company began back in 1870, three years before my father was born, my grand-father Hiram, known in the Lumber trade as "H.O.", set up a saw mill on a mountain stream near his home, to saw Spruce building lumber.

Of course, water power was all the power they had in those days, and for many years afterward. The streams were small, and it is said that to run this first mill, water dropped about 80 feet thru an 18-inch penstock. But the power was good at times, and when they could not saw, they cut logs, which were right near the mill in those good old days, and big ones, too.

Of course, there was no competition from Western lumber, and as the country from here to Boston was growing rapidly, there was no problem in selling the production at about $12 per M, loaded on the cars.

Another big item, was Spruce clap-boards, for they were the only siding available, and every wooden building had to use them. They were a very profitable item, since they were sold as inch lumber, while the thickness went from ½" down to nothing. We still make them, and as far as I know, we are one of two companies that still manufacture quarter-sawn Spruce clap-boards on the same model of machines that have been used for a hundred years. The real reason why they are no longer made to any extent, is of course, the availability of western clap-boards, re-sawn, and the virtual impossibility of getting any large quantity of old growth clear spruce logs, from which they are made.

About 1890, grand-father moved to Moretown on the Mad River, where he bought out two mills and power sites, about a third of a mile apart.

One was a saw mill, and the other a clap-board and grist mill, combined.

Even when I was a small boy, some farmers raised their own grain and had it ground at the grist Mill. In combination with the Saw Mill, grandfather set up a box shook factory. Wooden boxes were a profitable item, since, as you can imagine, tremendous quantities were in demand as paper boxes were unheard of. They furnished an outlet for the poorer grades of lumber. I used to work in the box shop when I was a kid, for $1 a day.

The lumber to make them was sawn round edge, and stuck to dry. Then it was drawn to the Mill, and the first, operation was to put it thru a one-sided log bed planer. I don't remember as it was ever called a single surfacer in those days. Then we threw the boards back along side, and ran them thru, again to 13/16" thickness. Then they were cut to length on a swing saw, and edged on a sliding table by a rip-saw. It did not make much difference if they were wider at one end than at the other. The pieces were then put thru a vertical roll matcher, one side for the tongue and the back for the groove. After that, they were driven together on a table with a big wooden mallet. Finally, the whole piece, which was either the side, top, bottom, or end, was passed on a table by rip-saw to size them. The ends usually had to be cleated. Thus, a shipment of knocked down boxes went to the purchaser, ready to be put together.

Boxes were about the first good use for Hemlock lumber. Before this, Hemlock Lumber was drawn to the cars by horses, and sold for $8–$10, per M.

Hemlock bark, however, had a big sale for tanning leather, before chemicals were used, and many times, giant Hemlocks were cut down, the bark peeled off, and the logs left in the Woods to rot, which took many years, since there is none of our wood that is more rot resistant than Hemlock. It was used for years, and still is used for planking on wooden bridges.

H.O. Ward died, in 1912. I believe he was smarter than we have been since, for he put all his money back into buying mountain timber, which in those days did not tie up much money.

He was the first man I ever heard of who had an ulcer, which is such a hazzard among businessmen now-a-days, and altho he went out west to the Mayo clinic, and tried all kinds of cures, it finally took him away. He worked in the Mills a lot himself. But, he always wore a black derby hat to top off his six-foot frame, making him an awesome, figure in my eyes.

He worked hard, and the government let him keep what he earned, and in that respect was much more fortunate than are we, and I suspect he had more real enjoyment, business-wise, than we. He must have, for be worked every night in his office.

But I must hurry on.

Hard-woods were coming into the picture during the last of H.O's life, when my father, Burton, took over the business. One big outlet was a plant making wooden window screens and doors, in Winooski, Vt. We sold them great quantities of medium grade hardwood over many years. The old time wooden box business petered out, and in that mill, we began to make hardwood chair stock, now called furniture dimension. This utilized our large edgings, sidings, etc. The price paid for it was low, but so were wages then. It was a marginal operation, and maybe it still is.

Now, father, Burton, became very much interested in "Reforestation," and about 1915, began to plant trees in the meadows of farms we bought for lumber. We have planted some every year, since, almost 2,000,000 seedlings, largely spruce and pine. Now, however, with the increase in wages, taxes, etc., I do not believe it is economically feasible to plant trees, for plantations should be tended, that is, pruned, thinned, weed trees cut out, etc.

On the morning of Nov. 3, 1927, Burton and my brother Kenneth went out of our valley to look at some timberland. Before they could get back, the big flood struck, washing out every bridge in town. The Mad river was really mad, and we watched over a million feet of our lumber float toward Lake Champlain. It was impossible to salvage much because it was covered with grit, making finishing impossible. However, we survived that and also the depression which lasted about 10 years.

In 1935 our large mill burned, and at that time we were able to buy a complete mill out-fit, including the six-foot band saw we now use and that has served us well.

My brother Kenneth died just as World War II was starting, and father and I carried on. We did not make as much money during the War as some people did. I guess we were not smart, but Ward Lumber Co. was fairly well known, and if any government men were around checking prices, they made sure we were in line, not that we wanted to do anything else of course.

After the War, things went along fairly well. My father spent most of his time working in his plantations. He would take a group of teenagers to the Woods and give them so much a tree to prune up about 8'. This cannot be done at present, due to child labor and wage hour laws. Now the teenagers hang around and do nothing good. If they were allowed to work they might cut themselves on a saw. Burton Ward died suddenly, at 79 years old, in the woods, after having worked, all day pruning trees.

During the 1950's, two more Ward's came into the business, Owen, my brother Kenneth's son, and my son, Holly. My oldest son, Richard, is a wholesale Lumber Salesman, Owen and Holly have done a lot of figuring, and about the first change was to discontinue sawing suit wood. In our hard-wood Mill, we were one of the first in New England to install an automatic carriage. It is compressed, air-operated, and we have had very good service from it. About the same time, we put in a live log deck, steam niggers and an long edge sorter.

These devices were a good start toward automation, but our location in Moretown was poor, and also, the old mill building just wouldn't take it. Things were always going out of line, and, we still had too many lines of shafting belts, etc. Of course, years ago, Mills had to be located by the river because of the water power. We had good water power, and used it to some extent up to a couple of years ago. However, it is too slow for modern machines. Now-a-days, we must get instant, fast, production, and not have men waiting for the Mill pond to fill up, or for leaves and other debris to be raked off the rack. The dampness and fog by the river was also a problem from the standpoint of Lumber drying. We had to get away from it.

Last year we had the chance to buy the Plant of a Plywood Co. of Waterbury, nine miles away. The building contained 44,000 sq. feet of floor space, equipped with a sprinkler system, on the rail-road, adjacent to the Interstate Highway "89", with a large flat field. We were surely sorry to leave Moretown after 70 years, but we bought this property, and it is ideal to our operation. We have more space than we need to install our Saw-mill and Dimension Mill, leaving plenty of space for storage and a small pre-drying operation.

Our saw-mill is on the upper deck, as is the De-barker for we produce about 20 tons of de-barked slabs per day. These fall to platforms on the ground floor, as do the edgings and sawn Lumber. The Lumber is piled on wagons and drawn to the edge sorter, which is located away from the Mill. Thus we are able to sort Lumber into piles whenever we want to, and also, sort lumber that we might buy from others. The slabs are bound up, and placed on trailers which are drawn each day to a chipping plant. The Mill is so arranged that the Sawyer, edge-man, and chipper man are all that are needed on the floor, and because of surge points at separate places, where round edge and square edge Lumber accumulates, on decks controlled by the Sawyer, he alone can saw at least 5,000 feet, before edging and clipping is necessary. Of course, a fan below has to take care of the slabs, and when edging and clipping is going on, two other men have to work below, piling lumber on carts, and working the edgings over.

Also, on the ground floor, is the Dimension Mill, beginning with a Bolter for short logs and ending at a turn-table. We believe this operation is O.K., but it is full of problems.

I was asked for my views on future developments. I realize that future views are only popular if they are optimistic. I am by nature, optimistic, but I must say that I am not too optimistic to the chance of large profits from regular lumber manufacturing in this area. The reasons are these, as I see them:

1. This area is running out of good logs. O.K., I know people will say, "we have been hearing that for years", and production is right up there. That is true, but it has been held up there because mechanized logging equipment has made it possible for Timber to be taken out from places where horse logging could not operate.

2. The farmers who lived in the hills and cut logs each winter, are gone. They have been squeezed from their farms after they stayed just as long as they could by cutting off all their marketable Timber. The lights we used to see at evening on mountain sides, have gone out, and the farmer and his boy have departed to the City to swell the unemployment rolls. And now, Nature is moving in rapidly to close the gaps by growing hap-hazzard variety of trees. Thus it is said that Timber is growing faster than we can cut it, and no doubt that is right, but 75% of it is far from marketable at present. Of course, the giant Birches, Maples, Spruce and Pine, will never be allowed to stand again. It is, however, the duty of every woodland owner to spend as much time as possible in his woods, cutting out poor quality, and so-called weed trees. This I believe is more important than planting, and will be more profitable.

It boils down to this, as we see it; with smaller diameter logs, higher labor and maintenance costs, all the automation possible must be the answer for the Mill operator, aided of course by the sale of by-products like de-barked slabs and saw-dust.

This is what we are trying to accomplish. In addition to this, we have had our timberland professionally appraised, to see how much timber we have in various stages of growth. From this appraisals, it looks as though we could cut, over 3,000,000 feet off our own Land each year, and after 10 years, due to cutting trees, have a better stand, that have reached their maximum value

We have operated this family business for a great many years; and have no intention of cutting all our timber and moving out. We have not set up our Mill with big production in mind, but we would like to saw about 15,000 feet per day, with just as small expense as possible. This Saw mill, with our modern Hard-wood Dimension set up, is about all we want to look after at present. It should give our three families a fair income, and probably we will have to cut Uncle Sam in on a little.

As I review the things I have said to you, I am dismayed that there is so much family reference, then I am compelled to realize a presentation by any of this group would have to be that way. We, in any Lumber program, are just a little different from most business development. No one of us suddenly creates a chemical formula that revolutionizes the world; that can save or destroy it; we do not dig underground for precious minerals, gems or fuel, but our raw material grows in slow, majestic splendor, offering a panorama of color and change: from Season to Season. In spite of the frantic pace of the world, we still wait upon Nature. But as we wait, we need to be practical; we have families, and families to come, who, we hope, may choose this "way of life."

Thru organization such as this, we can share our ideas, profit by the mistakes and success of each other, as we explore and "utilize our integrated product."

I close with this thought from Matthew Arnold's:

Yes, while on earth a thousand discords sing,
Man's fitful uproar mingling with his toil,
Still, nature's ministers move on
Laborers that shall not fail, when man has gone.

Merlin B. Ward President
Ward Lumber Co., Inc.
Waterbury, Vermont

Editor's Note: The above speech is reproduced as it was written, and presented, word for word, with the spelling, punctuation, grammar and terminology in the original text.

1968 — SOLD

In the 1960s, Vermont's population was growing and the state was changing. The Interstate highways made formerly remote areas accessible, skiing was soaring in popularity, and new industries—such as IBM in Essex—were driving the state's economy in new directions. With more people living in and visiting Vermont, property values in many parts of the state were rising.

For decades, the Ward family and lumber company owned thousands of acres of land—more than 30,000 altogether. This working landscape was slowly producing crops of trees. With trees' slow rate of growth, parcels of land might be selectively cut every 25 years or so. Between those harvests, the acreage was not producing income.

As land values rose in Vermont, property taxes were increasing, too. The tax structure at the time did not take the use of land into account in its appraisals. A parcel of land that was a scenic house site and a parcel next door that was owned as a source of logs were taxed the same. Vermont's Use Value Appraisal Program, the Current Use law, was not adopted until 1978. As the Vermont Department of Taxes explains, "The purpose of the law was to allow the valuation and taxation of farm and forest land based on its remaining in agricultural or forest use instead of its value in the market place. The primary objectives of the program were to keep Vermont's agricultural and forest land in production, help slow the development of these lands, and achieve greater equity in property taxation on undeveloped land."

In the 1960s, as the Wards looked at their landholdings, this law was still far in the future.

"You couldn't justify paying these taxes on lumber that would take decades to grow," explains Owen Ward looking back. "We didn't know that the Current Use tax was coming."

The value of choice pieces of land had appreciated considerably. Buyers were interested in accessible scenic parcels. "We were always being asked to sell this piece here or that piece there, but then you had to decide about selling road frontage and where it left your other land," recalls Owen. "We were not a real estate development company and did not see ourselves as real estate developers."

Through these years, the mill in Waterbury was running smoothly, but the statewide changes in land use and land value impacted the company. In the 1960s, the Wards began assessing their holdings and started exploring changes of their own.

TIMES-ARGUS, BARRE-MONTPELIER, VT.,

Ward Land Sale Is Completed

WATERBURY — The Ward Lumber Co., a fourth-generation Vermont business, and members of the Ward family, yesterday completed the sale of 30,000 acres of timberland in central and northern Vermont.

Laird Properties Inc., a subsidiary of Laird Inc., a New York investment banking firm, which made the acquisition, becomes one of the largest landowners in the state.

Involved in the property transfer are 53 separate parcels of land in 15 towns.

These towns include Moretown, Middlesex, Fayston, Waitsfield, Warren, Duxbury, Northfield, Waterbury, Worcester, Calais, Granville, Williamstown, Bolton and Victory.

The principals in the Ward Lumber Co. Inc. are Merlin B., Mrs. Aline H., Owen M., Richard S. and Holly M. Ward and Mrs. Lois W. Tierney.

Burton Ward started the lumber business in Moretown. He was recognized for his efforts in

firm in New York, represented the new owners.

Included in the acquisition is the Ward Lumber Co. mill in Waterbury. The new owners will continue its operation.

The president, Arthur Collins, of the New York corporation, said the purchase was made primarily as a long-term investment and that there are no early development plans.

He added, "We are well aware that the character of the development of any property as large as ours will have a serious impact on the character of this entire section of the state.

"As landholders for the long term, we share the concern and interest of the state and the area residents in maintaining high standards for the use of this crucial property.

"Over the course of the next year, we will be approaching the interested authorities in Vermont and the best available land-planning talent to insure that future development

December 31, 1968
Barre-Montpelier Times-Argus

After negotiations, on December 31, 1968, members of the Ward family sold Ward Lumber Company, including the Waterbury facility and about 28,000 acres to Laird Properties, a subsidiary of a New York investment banking firm. The property included 53 parcels of land in 15 towns. Ward family members retained some acreage and the Ward Clapboard Mill in Moretown village.

Leo Laferriere was working for Laird at the Waterbury mill after the sale and remembers a conversation with Owen after finding Owen's *1923 Scribner's Lumber & Log Book* in Owen's former desk drawer, the desk Leo was using. No one in the lumber industry would be caught without their copy, as it was invaluable. Owen remarked, "That used to be a very valuable book." Leo agreed, but added: "It still is valuable. You see, Owen, right inside, on the cover, you've written out, for all to see, the combination to the company safe!"

In fact, to this day, there are three combinations to safes in that little book. One on the back inside cover and two on folded pieces of paper tucked inside. The safes are long gone.

Owen's pocket-size Scribner's Lumber & Log Book

Owen Miles Ward

Born in 1925, Owen Ward worked with the family company from childhood to its sale in 1968—with some time away for his education and military service during World War II.

Kenneth Ward's older son, Owen planted and pruned trees on Ward plantations when he was a boy—and saw those same trees reach the mills decades later. A 1943 graduate of the New Hampton School in New Hampshire, Owen attended Syracuse University where he was a member of Phi Gamma Delta. He served in the U.S. Army Air Force during WWII flying in B-29 bombers.

Through the 1950s and 1960s Owen led numerous innovations in the business; several of these greatly improved operation of the Lower Mill. These included investment in equipment such as the automatic log carriage, a powerful turning lever that flipped and rotated logs as they were fed to the saw, allowing more efficient use of the logs; an electric-powered hog that ground slabs into wood chips that were then sent to the boiler room as fuel. As bark has more BTUs than the other wood, this was especially cost effective. The addition of the automatic edge sorter brought even more efficiency. This entire system, along with other mill equipment moved to the Waterbury facility in 1962.

The two-generation team of Merlin, Owen, and Holly Ward were at the helm of the company for about two decades. These years saw considerable changes in the industry as new technology became available. Merlin, in his speech delivered to the Northeast Logger Congress in 1963, credited Owen and Holly with leading these initiatives that automated the mills, moving the business ahead during this time.

Owen in the Air Force

All three Wards were involved with the company's landholdings, and during the 1950s and 1960s land use in Vermont was changing. One transaction that Owen helped negotiate was the sale of company lands that became Glen Ellen (now Sugarbush Mt. Ellen) ski area.

Beyond his work with Ward Lumber, Owen served on the board of the Chittenden Trust Company. Serving for many years on the board of directors of the New England Forestry Council, he was involved in formulating recommendations for promoting good forestry practices.

Owen was a member of the Mad River Lodge #77 F.&A. Masons of Vermont for over 50 years and a longtime member of the Moretown Methodist Church. He maintained ties to the church and community even after the sale of company.

To honor their father's memory, Owen and his brother Wyman donated to the State of Vermont a beautiful parcel of Moretown land at a well-loved swim hole on the Mad River. This public property, the Kenneth H. Ward Memorial Access Area, is enjoyed by thousands of people every year.

Owen's high school photo

1932

At 7

Owen Ward in front of the Moretown clapboard Mill, June 2011

Owen, Wyman and Florence

Holly Merlin Ward

Today, at least 140 years after Hiram Ward cut his first clapboards on Dowsville Brook, Holly Ward continues this tradition. At his small mill in Moretown and a modern mill in Maine, he produces vertical-sawn clapboards—his Maine mill also cuts lumber for sounding boards for musical instruments. In Moretown, these high-quality boards are still cut with the company's historic lathe and saw—machines that date back to the early 1900s or even late 1800s. The Maine mill's equipment belongs to the modern era.

Born in Moretown in 1930, Holly was attuned to and interested in the lumber business from childhood. Like his siblings and cousins, he planted and pruned trees in the plantations and helped out with a range of company-related tasks. His experience with the company's big work horses contributed to his lifelong equestrian interests. After earning his student pilot's license as a young teenager at a local air field, Holly prevailed on his grandfather Burton to go up with an experienced pilot for his first flight ever, an opportunity to get an aerial perspective on Ward lands.

Holly graduated from Montpelier High School and served in the U.S. Army, and Army Reserves. After two years at Vermont College, he continued his studies at Boston University, graduating in 1952 with his degree in industrial management. When he returned to Moretown, he joined his father, Merlin, and his cousin, Owen, in the family business.

Beyond overall involvement in the company, during the Moretown years, Holly's focus was generally the Upper Mill—a mill that saw considerable advances during his years, including expansion into the building on the uphill side of Route 100B.

The softwood operation continued there, and the addition of the bolt mill allowed the company to be more productive with available hardwood timber supplies. With the bolter and the Porter Rough Mill System, Ward Lumber served changing markets and produced an astonishing quantity of high-quality parts for new furniture markets of the 1950s. The Porter system equipment was lost in the 1962 Upper Mill fire. With the company's move to Waterbury, the bolt mill moved, too, and was finally united in a single facility with the systems and equipment of the Lower Mill. This consolidation allowed the company to efficiently work with both high and low grade logs.

True to family tradition, Holly was, a longtime member of the Moretown Methodist Church. With the Ward Clapboard Mill, he continues to be involved in the Moretown community.

Holly ready for the slopes

Summertime with friends

Lois, Holly and Richard

The Fourth Generation Siblings

Richard Smith Ward

Merlin and Aline's first son was born in 1924. After graduating from St. Johnsbury Academy in 1940, Richard began his studies at Syracuse University. Soon after the United States entered into World War II, Richard convinced his father to sign a waiver allowing him to enlist in the military at age 17.

Richard served in the 1st Marine Division, the oldest, largest, and most decorated Division in the United States Marine Corps. He served through the duration of the war, making five landings in the South Pacific including Guadalcanal, Cape Gloucester (twice), Peleliu, and Okinawa.

After his distinguished military service Richard became a wholesale lumber dealer, establishing his own successful business based in Vermont.

Lois Evans Ward Tierney

Born to Aline and Merlin as the 1927 floodwaters receded, Lois was the one daughter of the third generation of Moretown Wards. Lois, like her brothers and cousins, grew up with the family business.

During World War II, as a teenager, Lois did her part for the war effort. She was one of the community members who volunteered as a plane spotter. Local plane spotters worked shifts in a small shack-like station in Duxbury watching the sky for aircraft. Trained to recognize different types of planes, they were watching for foreign aircraft and assisting in the defense of the U.S. northern border.

Lois graduated from the Northfield School for Girls in Massachusetts in 1945 and Syracuse University in 1949. She married Robert Arthur Tierney of Long Beach, Long Island, June 17, 1950.

Wyman Burton Ward

Kenneth and Florence's younger son, Wyman was born in 1935. After his childhood in Moretown, Wyman graduated from the New Hampton School and Bryant College. Serving in the U.S. Army, he was based at Fort Hood. Wyman built his career in business administration in Rhode Island and Connecticut, including serving at vice president of Chandler-Evans Manufacturing in Hartford.

Richard, Lois, Holly, Wyman and Owen Ward

The Clapboard Mill Today, as Yesterday

When Hiram relocated to Moretown from Dowsville Brook, the clapboard mill moved, too. Today it is still there, just upstream and uphill of the site of the former Lower Mill. The mill's original equipment dates from the late 1800s. After the sale of the Ward Lumber Company, the Wards retained the ownership of this mill.

Today, under Holly Ward's direction, the mill still produces high-quality quarter-sawn or vertical-grain clapboards cut from clear pine and spruce. Water power ran the machinery and its connected shafts, belts, and bearings until electricity replaced it. The process is the still the same.

Logs are delivered and rolled down a ramp to the entrance of the sawmill. Once inside the mill, a center-point is marked at each end with a metal template. Two bars with hook-ends are attached to the ends of the log to lift it and swing it into place in the lathe where it is centered using the template markings, and clamped into place. Once in position, the log spins, a sharp V-shaped blade travels back and forth along the rotating log, shaving off its bark and bringing it down to an even diameter along its entire length. Within minutes, the log is an evenly sized cylinder. Back come the lift chain and hook, and the log is swung around to the saw where it is clamped into the carriage. As the saw's circular blade spins, the carriage lowers the log, is adjusted for the correct depth for that log, and then carries the log back and forth over the blade. After that first cut is complete, the log is rotated a precise few degrees to start the next cut. With the rotation and repeated passes, the entire log is cut in thin wedge-shaped slices. The narrow end of each board is still attached to the heart of the log; it is then simply snapped off. The remaining thin cylinder of wood, the heart, may become a fence post or biomass fuel.

After the individual clapboards are snapped off, they are stickered, set in stacks with thin strips of wood between the layers to allow air circulation. After they have dried, they run through the planer, a machine that shaves their surfaces smooth. Then they are ready for sale.

Clapboard width is measured in half-inch increments. Ward Clapboard Mill cuts them in widths from 4.5" to 6.5". The circumference and diameter of the log determine the number of clapboards that can be cut from it—a 15" diameter log will yield about 60 clapboards. Each individual clapboard is $9/16$" at its wide end. By cutting the logs this way, the growth rings are perpendicular to the face of the board. They are quarter-sawn clapboards of higher quality and more expensive than their flat-sawn peers.

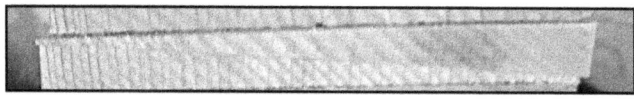
Growth rings perpendicular to the face of the board

Looked at in profile, the rings of quarter-sawn clapboards line up vertically. These clapboards are less likely to crack or pop nails when hammered into them and they are dimensionally more stable than other cuts of clapboard.

Clear logs are essential for high-quality clapboards. Lower-quality logs produce clapboards that have too many defects in them to be useful. Without knots, a clapboard can be six feet long.

In the 21st century, the Ward Clapboard Mill in Moretown, and its modern facility in Maine, continue the Ward tradition.

Above: a recently cut pine from one of the former Ward plantations awaiting entry into the clapboard mill.

Middle: rolled inside the mill in front of the lathe, the log will be marked with its end center points, then hoisted into the lathe.

Below Left: Bob Pierce uses a template to mark the center-point of the log.

Below Right: Bob is marking the other side. Notice the × Marking the center point on this side of the log.

The process of creating clapboards from logs, still using the original equipment in Moretown

Above: Bob and Holly loading the log into the lathe where it will have the bark removed and the entire log will be brought down to an even, consistent diameter.

The lathe's blade slides back and forth along the rotating log until the log is down to an even diameter, all bark removed.

Bob then marks the log for its later use with blue crayon, while the log is still spinning. This leaves the marks on the ends of the clapboards once they are cut and removed.

125

Above: Using the hooks on the chain again, Bob swings the log over to the saw (to the right of Holly Ward).

Below: Bob loads the log into the saw's framework making sure the center point is exactly the same as it was on the lathe.

The saw in action, just before the log is rotated to start the 7th cut. The carriage slides back and forth allowing the blade to make each cut in and out. The log is then rotated slightly for the next cut.

Above: Bob snaps off the sawn clapboards.

Below: Because this part of the tree was the upper portion of the trunk, it was above the area pruned years ago, now showing knots, the legacy of branches.

Sorting the clapboards before stacking them to dry.

Holly is starting a drying stack making sure there is space between the boards to allow air to flow around them.

Below: the planer with 2 boards running through, smoothing them out in the final step before shipping to market.

The tradition continues...

The Moretown mill today is housed in these buildings just above the location of the old lower mill. The little building at the right in the background houses the sawmill where the clapboards are cut. The building on the left houses the drying clapboards and the planer. The finished clapboards are then stacked within the open door of the building awaiting stacking on trucks for delivery. The middle building is a shed for "outside" equipment, and to the right, a chute is located on the roof into which the sawdust is blown from both buildings (the pipes between the buildings) to be loaded onto a truck for delivery to farmers, horse farms, and others purchasing sawdust. Some of the product becomes wood pellets for stoves.

Ken Ward, Bob Pierce, Owen Ward and Holly Ward at the old clapboard mill, June 2011

Ward Lumber Company Timeline

1842
Hiram Owen Ward born Duxbury 1/10/1842

1872
H.O. Ward operating sawmill on Dowsville Brook in Duxbury.

1874
H.O. Ward diary (see following pages)

1884
H.O. Ward adds clapboard mill in Duxbury.

1888
H.O. Ward builds mill in Moretown at site of burned grist mill.

1889
H.O. and May Ward move to Moretown. Ward mill there employed 6 men and produced 1,000,000 feet of lumber annually. In 1890s, ships 17 boxcars of lumber per month from Middlesex.

1905 (circa)
Wards buy general store in Moretown.

1910 (circa)
H.O. Ward and Burton Ward start tree plantations

1914
H.O. Ward dies, May 9. Estate includes thousands of acres of forest land and abandoned farms.

1927
Vermont Flood of 1927, extensive damage to Moretown, mills, and dams.

1935
Lower Mill burns. Rebuilt that year.

1938
Flood and damage and loss of lumber from New England Hurricane.

1955 (circa)
Wards buy the other store in Moretown, opposite the Town Hall. This becomes the Ward IGA (grocery store.) The other becomes the Ward Hardware Store.

1955
Fire destroys Upper Mill—sawmill and chair stock mill in Moretown.

1956
Second Upper mill built up on the storage area (building still exists as a home).

1961
February 2—Fire again destroys Upper Mill along river.

1962
Sale of Owen Ward/Paul Bigelow land to Walton Elliott for Glen Ellen ski area.

1962
June—Ward Lumber Company buys Pilgrim Plywood Corporation plant on Railroad Street in Waterbury. Mill operations move from Moretown to Waterbury.

1964
April—Hardware Store building burns.

1967
May—Wards sell Moretown IGA Store.

1968
Ward Lumber Company, including 29,000 acres and Waterbury mill, sells to Laird Properties, Inc., of New York.

Early Moretown at the Time of H.O. Ward

From across the river looking down on early Moretown

Courtesy of the Moretown Historical Society

An early view of Main Street, Moretown

Courtesy of the Moretown Historical Society

The Diary of Hiram O. Ward, 1874

January

Thursday 1–go to Waterbury with lumber and get meal. Frenchman killed at Waterbury today Asa Corliss took horses to Dowsville and I came home.

Friday 2–I chore and go to Moretown with Thomas Neill. Go to Corey place and see about Asa's bank p.m.–Newton drives horses and Asa oxen.

Saturday 3–Mr. Somerville came and killed last beef a.m. I visit at Somerville's with Mr Holmes people in afternoon.

Sunday 4–Meeting at our church, small attendance. [thawy day?] Aunt Dency quite ill.

Monday 5–Go to mill and stick up lumber–and pile logs, etc–Newton does not work.

Tuesday 6–Draw 2 loads logs a.m.–Stormy, do not work p.m.

Wednesday 7–Come home early. Ed Belding goes up to fix aqueduct. I got sand and have plastering done in S room.

Thursday 8–Carry Frank and Clinton to see Dr. Haylett. Frank has tooth drawn. Tom splits wood a little p.m.

Friday 9–I go to mill and help Canerdy saw. Saw 4200 feet. Newton works a.m.

Saturday 10–Help Canerdy saw. Wm [William] fixes road and saws two loads logs.

Sunday 11–No meeting at our ch[urch]. Mary and I attend at Moretown p.m. Very few present.

Monday 12–go to mill and get load hwd [hardwood] and draw to Waterbury. Come home at night.

Tuesday 13–May, Clinton and I go to Waterbury, settle with Wyman and Smith, ec [etc].

Wednesday 14–Go to Moretown, settle with Geldings and Blacksmiths. Get census records, ec [etc]. A little snow.

Thursday 15–Go to Dowsville, [Grgn] then to measure hwd lumber for winger co.

Friday 16–Go to Dowsville–measure remainder of pile lumber. Draw load to top of hill near Wm morning. Tom splits wood p.m. Mrs. Seaver buried today.

Saturday 17–go to Dowsville for lumber–set job of drawing lumber to Stock well. Tom splits wood some. I draw lumber to Waterbury and get meal, ec.

Sunday 18–meeting out our ch. Good attendance and good Sabbath School. May unwell did not go to ch. Mrs. Folsom played.

Monday 19–Rainy day. I go to Wms a.m. and to Moretown p.m. sell hide and settle with Wyeth & Co.

Tuesday 20–Go to mill. Stick up boards–a.m. Work in woods p.m. Tom drew boards from mill and stuck them up p.m.

Wednesday 21–Went to Crans [?] a.m. Tom hires out to Green [?]. A little snow today. I went to Moretown and to Tum?? Hill p.m.

Thursday 22–Tom goes to Granville. I draw lumber to Waterbury. Mrs. Smith and children came today.

Friday 23–Massive [?] lumber at mill. Newton helps K handle it. Draw load to Com?? at night.

Saturday 24–I go to mill, get load lumber and draw to Waterbury, get flour &ec.

Sunday 25–very cold day. No meeting at our ch–

Editor's Note: Many abbreviations are used throughout the text, the first few times used are indicated in brackets. Quite a bit of text is unreadable. In some places written over, or illegible. Where it made sense, a best guess was made followed by [?].

attend funeral of Mrs. Clough and visit Aunt Dency p.m.

Monday 26–very cold. Do but little aside from chores.

Tuesday, 27–attend funeral of Mr. Pease's child. Ed Smith home for Mrs. Smith p.m.

Wednesday, 28–Draw hay to mill and take load boards to Comer [?] at night. Snow storm p.m.

Thursday, 29–Go to mill a.m. & get horse shod. Carry Clinton to do hooker and visit at Mr. Holmes in afternoon.

Friday 30–Go to Dowsville and draw logs–Take load to Comm at eve

Saturday 31–Very stormy remain at home. Sleighing again.

February

Sunday 1–meeting at our ch. Pretty good attendance and good Sabbath School.

Monday 2–thermometer 42 below zero. Go to Waterbury with lumber and get meal & flour +c. draw it to Dowsville and come home at night.

Tuesday 3–Take hog to Moretown and split wood a little p.m. Chester came at night.

Wednesday 4–visit with Chester a.m. Go to Wms with him p.m. Split wood some afterward.

Thursday 5–To go mill and work at waterwheel some a.m. Help Corliss on logs at dump p.m.

Friday 6–White wash kitchen and split wood some in afternoon. Cold weather.

Saturday 7–Go to Dowsville & draw logs – Newton does not work

Sunday 8–no meeting our ch. Remain home all day.

Monday 9–go to Moretown a.m.–draw logs p.m. & get horses shod at evening

Tuesday 10–Draw logs–get along well. Newton drives oxen.

Wednesday 11–draw logs a.m. Nat Thomas commenced to work at noon at $23.00 per month. While he works–he is to quit anytime wish and I am to pay him for the days he works only. I pile boards &c p.m.

Thursday 12–Pero fixes barn door. I chore a little and draw load wood from Mr. Wheeler that Asa Corliss got out.

Friday 13–Pero goes to Waitsfield, I go to Waterbury with Mr Somerville's people get money of Fred Smith, see Mr. Wrigley &c–Thawy snow all gone [meal at night?]

Saturday 14–thawy day. I go to Chipman & have ox attended to go to Moretown &c–Mr. Gay & wife here on visit p.m.–Newtown helps Canerdy saw. Wm & Shawn don't work.

Sunday 15–Meeting at our ch. No sleighing & not very large attendance.

Monday 16–I went to Dowsville. Mr. Wrisley then to look over mill. Pero helps paper a little in afternoon.

Tuesday 17–Saw shingles stuff a little a.m. Visit at S. M. Indus in afternoon, have Mr. Kellogg's horse. Cold at night.

Wednesday 18–Chore a little a.m. Mr. Stevens here to sell milk. Plans to attend oyster supper at Mr Kellogg's pm.

Thursday 19–Wm here for hay am. Teams draw logs pm. I chore &c am attend funeral of Uncle Hezekiah Ward pm. Came home eve.

Friday 20–Horse teams came home am–I get shed shod–pm. Pero makes shingles some.

Saturday 21–Go to mill and get horse shod &c Williams ox lame, he takes them home.

Sunday 22–Attend ch at Methodist Waitsfield am & funeral of Parson Palmer in afternoon.

Monday 23–Go to Dowsville at noon. Thomas draws logs alone with onex. Wm & I draw with horses in afternoon.

Tuesday 24–Draw logs at Dowsville. Wm away to get horses shod half day. Wilbur works with oxen.

Wednesday 25–Pero drives horses & I work in woods with ox teams–Wilbur & Thomas work.

Thursday 26–I work in woods awhile & then come home for meat & apple. Pero drives horses.

Friday 27–Draw logs and help on teamsters–Pero works also

Saturday 28–Pero and I work at logs also Thomas and Wilbur.

MARCH

Sunday 1–Meeting at our ch, fair attendance

Monday 2–Draw hay oats meat and potatoes to Dowsville and draw 5 loads logs to mill. Pero works. come home at eve.

Tuesday 3–March meeting day, Asa Corliss commences to work, he washes & scatters tubs – Teams finish drawing logs today & come home at night. I tap a few trees. Rainy at night.

Wednesday 4–Rainy day. Gather and boil a little sap. Pero shoes sled and draws hay a from ox barn to cow barn. Sugar off in evening.

Thursday 5–I went to Moretown with shingles & to mill &c. boys draw load hay to horse barn and fix tubs at sugar house–Pero here to sell milk Pans[?]

Friday 6–I went to Montpelier & Northfield with Wm to see about clapboard mill. Frank Pero split wood on hill. Asa drew manure. Cold in morning.

Saturday 7–Asa helped Wm with oxen Pero & I went to mill & drew many boards &c. Perkins of Montpelier then to see about putting up clapboard mill.

Sunday 8–Remain at home. Snowy most of day. Asa and Pero away.

Monday 9–Asa draws manure. Pero works at mill. I work at mill am & go to Middlesex with lumber in afternoon.

Tuesday 10–Wm and Pero draw boards to Middlesex. Asa drew manure. I went to Moretown and get district statistics.

Wednesday 11–Asa does not work. Pero draws manure. I draw hwd to Colby Co. Wm draws also.

Thursday 12–Asa does not work. Pero draws manure. I draw hwd to Colbyville & get load spruce to Comm…

Friday 13–I draw load spruce to Middlesex and skid hwd logs on Munson[?] place pm. Pero draws manure some.

Saturday 14–Draw hwd to Colbyville. Pero draws manure. Asa does not work.

Sunday 15–Last meeting at our ch for the present. Fair attendance.

Monday 16–Wm & I draw flooring to Middlesex– Asa does not work.

Tuesday 17–Pero & I go Dowsville, Wm there also. Put new sleepers on mill & also floor–stick up lumber some.

Wednesday 18–Wm & I draw lumber to Middlesex & go to Montpelier to see about clapboard mill – Pero helps Canerdy saw. Asa works in sugar place.

Thursday 19–Pero at saw mill, Asa in sugar place. I go to saw mill to get on logs & board off &c. Measure logs for Wilbur &c.

Friday 20–Pero at saw mill. Asa at sugar place. I go to saw mill & draw out & stick up boards & settle with Wilbur &c–draw down load at night.

Saturday 21–Pero at saw mill. Draw load boards to Middlesex & draw back freight for sawyer & Hills & Bulkley [?]. Very muddy going. Frank goes to mill with oxen. Asa works in sugar place, saw sap.

Sunday 22–Asa goes home. I remain at home. Not very pleasant.

Monday 23–William went to Montpelier. I went to Dowsville after boards &c. Canerdy came for force pump. Pero does not work.

Tuesday 24–Pero does not work. I go to Middlesex with boards–draw our XXXXX Bulkly & Hills.

Wednesday 25–Pero goes to Middlesex with boards. I sugar off &c. Asa draws wood to house.

Thursday 26–I work in sugar place in am and go to Middlesex with sugar in pm. Boys gather sap ec.

Friday 27–Boys boiled last night. Cow sick today. Sugar off –. Company Miss Price [?], Mrs Somerville and Mr Somerville.

Saturday 28–Asa watches cow and splits wood some. Pero does not work much – boil a little sap and sugar off once cool down [?]

Sunday 29–Pero at home. Attend the funeral of Mrs. Willard, Pastor of Fayston.

Monday 30–Pero & I go to Dowsville & stick & draw out boards. Draw down curb to Waterbury at night. Asa draws manure.

Tuesday 31–I go to Montpelier early. Get re????? done on waterwheel &c buy flour &c. Pero does not work. Asa works at manure.

APRIL

Wednesday 1–I go to Dowsville & sell lumber to Green +c go to Waterbury for belts and meal. Pero does not work. Asa skins dead cow and kills calf.

Thursday 2–I go to Dowsville, draw on logs, measure boards for Sull Shanw [?] & draw load hwd to Middlesex. Boys draw manure a little.

Friday 3–Pero & I go to mill & assort lumber &c. He draws shingle logs to his wife. I help board mill some, measure logs & sell[?] meal.

Saturday 4–Pero draws load of hwd to Middlesex. Asa draws manure a little. I work at Books and go to Moretown &c at night. Good wagoning.

Sunday 5–attend ch at Moretown with May & Clinton. Mr Bayer preached. Good attendance.

Monday 6–Go to Waitsfield & engage girl am. Pero and I go to mill & assort hwd am. Asa draws logs from hill.

Tuesday 7–Pero goes to Dowsville & helps Wm board mill. I go to Middlesex with lumber. Asa draws wood to house.

Wednesday 8–I sugar off. Asa gathers sap some. Pero goes to Middlesex with hwd. Good sap say. Miss Elizabeth Turner dead.

Thursday 9–I sugar off am. Attend to Wilbur & miner pm & boil sap some. Boys gather– Pero goes to mill for lumber pm.

Friday 10–Very snoy and windy. I go to funeral of Elisabeth Turner. Boys work at sugaring.

Saturday 11–Pero goes to Middlesex with hwd on sled & to Dowsville am. Mr Shumfeff here to buy clapboards. I sugar off XXX and boil sap some.

Sunday 12–Remain at home. Pleasant Day.

Monday 13–Draw 2 loads hwd to Middlesex. Wm draws one. Pero draws from mill with oxen. Asa does not work.

Tuesday 14–I go to saw mill with Wm & work at XX am &c. Pero draws load hwd to Middlesex & goes to Dowsville afterwards. Asa work at sap.

Wednesday 15–All in sugar place am. I go to Waitsfield, Fayston & Warren trying to find a clapboard sawyer. I asked there (in trouble)

Thursday 16–Sugar off twice & boil sap some. They start clapboard mill today.

Friday 17–I go to mill and work all day fix log way– get on log &c. Pero goes to Middlesex with hwd.

Saturday 18–Sugar off am. Go to Moretown pm. Boys work in sugar place. Sap runs well in afternoon. Boys draw hay pm some.

Sunday 19–Attend ch at Moretown am, also Clinton & May. Pero at home, also Asa.

Monday 20–go to Waitsfield for girl am. Sap plenty[?] has wasted a good deal.

Tuesday 21–Sugar off & work in sugar place. Boys work at wood some.

Wednesday 22–Carry May to Moretown am. Sugar off afterwards & go to mill & get on logs pm. Pero does not work–has horse to South Fayston. Asa works in sugar place.

Thursday 23–I go to mill and get hwd & draw to depot. Boys work in sugar place. Roads very bad today.

Friday 24–Stevens here to sell large pans. I attend to him in forenoon & sugar place a little in afternoon. Pero goes to mill for hwd & draws to Belding.

Saturday 25–Pero goes to Middlesex with hwd. Asa & I work in sugar place.

Sunday 26–Heavy fall of snow. Conference Sabbath & no meeting at Moretown. I remain at home all day.

Monday 27–Asa does not work. I go to mill & draw hwd with sled. Wm and Wheeler [?] also draw loads. Snow melts & we leave lumber near Rufus Holding. Pero goes to mill for lumber am and gathers later pm.

Tuesday 28–All work in sugar place some sap runs quite well.

Wednesday 29–Work in place–and watch with sick cow. Pero goes to Dowsville and gets 2 loads hwd boards pm.

Thursday 30–Pero work in sugar place am & goes to Middlesex with hwd pm. I sugar off & sell hay to Atturton & Somerville. Cold & snowy. Pero runs sled.

MAY

Friday 1–Pero goes to Middlesex. I sell hay all day. Asa works in sugar place. Frank watches cow.

Saturday 2–watched with cow last night. I sell 2 loads hay am–attend action at Waitsfield–buy buggy wagon. Boys work in sugar place.

Sunday 3–May, Sissi & I attend ch at Moretown. Mr Willis preached, feel very tired at night.

Monday 4–Pero, Wm, & Wheeler drew hwd to M... [Moretown?]. I sugar off & sell hay &c. pleasant day.

Tuesday 5–Boys work in sugar place. I weigh hay for Mr. Corsi and Bisbee am. Go to Moretown pm. Sell huli[?] by book. Sunny[?] &c pleasant.

Wednesday 6–Sugar off &c. Boys gather sap for beer & gather tubs. Tyler here to buy potatoes – Mary carried May & Clinton to Waterbury.

Thursday 7–Boys wash tubs &c. I go to Wilburs for hops am. Corsi here for hay. I go to mill and put additional posts made in am.

Friday 8–Make beer am. Draw hay to Dowsville and get sawdust in afternoon. Asa gathers spouts pm.

Saturday 9–Asa goes to Chipman and get oxen shod. Pero churns am. I weigh hay for Somerville &c am. Go to mill sell sugar &c am.

Sunday 10–Sissi & I attend ch at Moretown. Mr. Willis preaching.

Monday 11–Out beer in cellar and chored am. Go to mill for sawdust pm. Asa draws manure in afternoon.

Tuesday 12–Asa draws manure. I weigh hay for S.M. & S.U. Turner and unload sawdust &c. Jesse Corliss came at noon. He split wood pm.

Wednesday 13–Boys fix fence. I weigh hay for Somerville &c am. Go after plow & harrow & to find help pm.

Thursday 14–Go to mill for boards & harrow stuff am. Boys make fence on brook &c. I break up most of day. Pleasant. Go tomorrow at eve.

Friday 15–Jesse spreads manure. Asa draws. I plow on old sand beyond lower orchard. Abbi Taylor here.

Saturday 16–Settle with Wheeler for drawing bark[?]. Mr. Somerville gets hay am. Some rain. Boys move sugar, make posh &c I go to Moretown am.

Sunday 17–Sissie & I attend ch at Moretown pm. Pleasant day.

Monday 18–Asa does not work–I plow and weigh hay for Porter & Corsi. Jesse spreads manure some. Rain pm.

Tuesday 19–I go to Waterbury, settle with Smitleff & Wyman. See Grans (?) &–Boys draw manure & some fix fence & draw last of spouts and nails.

Wednesday 20–Finish plowing old land below sugar place. Weigh hay for Willin & boys draw and spread manure.

Thursday 21–Plow piece old land on Munson place & weigh hay for Crandall am. Go to Waitsfield & Moretown p.m. & attend wedding at Somerville in evening.

Friday 22–Boys work on fence on hill. I break up on hill p.m. Wilbur & M??? here a.m. (?)

Saturday 23–Boys churn and work on fence. I go to Dowsville for lumber and draw to Colbyville. Get flour meal etc...

Sunday 24–Sissi and I attend church Moretown pm. Mr Wallace a Scotchman preached Fairly [?]

Monday 25–Mr. Seaver came today. Assort and cut potatoes. Plant some ????. 9 bushels oats, Rainy pm.

Tuesday 26–Went to Moretown to Waitsfield for stump [?] & am. Joplin here and buys potatoes at am. Mr Seaver works in garden a.m. field pm.

Wednesday 27–Sow and harrow in hay seed. Mr Seaver plants potatoes. Asa draws manure. Break up some more for corn am.

Thursday 28–Finish breaking up for corn and plow south orchard. Warm day. Asa draws manure on hill. Seaver works at potatoes.

Friday 29–Sow oats in south orchard & on Manson place. Mr. Wheeler works. Asa draws manure on hill some & manures in hill for corn. Seaver plants potatoes.

Saturday 30–I plow in small orchard. Seaver & Wheeler plant potatoes and corn. Asa draws manure for corn. Sell hay to Wilbur.

Sunday 31–Sissi and I attend ch at Moretown pm.

June

Monday 1–Seaver horse came today. Seaver & Asa work at corn some [overwritten]. I go to Moretown to mill and get horses sod & harnesses repaired am. Go to Dowsville for clapboards pm.

Tuesday 2–Draw load clapboards to Middlesex am. Carry Sissi to Moretown pm. Seaver & Asa work at corn.

Wednesday 3–Draw clapboards to Middlesex. Asa & Seaver work at corn.

Thursday 4–Draw clapboards. Seaver & Asa at corn. Take dinner at Barretts.

Friday 5–Draw load clapboards to Middlesex & load lean [?] Asa works at fence & corn. Also Seaver

Saturday 6–Chore [?] a little am. Go to Dowsville for boards pm–Asa & Seaver finish planting corn tonight

Sunday 7–May and I attend ch at Moretown pm

Monday 8–Wm & I go to Ridleys with clapboards. Asa plows for Seavers potatoes. Small pox at Ridleys.

Tuesday 9–Draw load cbds [clapboards] to Middlesex & get corn–go to Waitsfield pm – have bad toothache

Wednesday 10–Sbb Ashly [?] . Asa works on road. Seaver plants potatoes. I work butter chore & visit with CH Ward and wife. Go to Crandalls at eve.

Thursday 11–Draw cbds to Depot & bring back flour, etc. Rainy at night. Asa plows price for fodder corn.

Friday 12–Wm & I go to Mill get logs on log way etc. Rainy nearly all day. Asa assorts potatoes & draws manure.

Saturday 13–Draw boards to depot am–sell corn to Wilbur, also flour. David Belding here buys pig etc. Asa sows fodder corn.

Sunday 14–May & I attend ch at Moretown pm. Presiding elder McAnn preached.

Monday 15–Mr. Seaver and Asa work on road with oxen. I draw load hwd to top of hill & load clapbds to corner at night.

Tuesday 16–Asa, Seaver & oxen on road. I go to Middlesex early–load corn.

Wednesday 17–Draw load hwd to Colbyville. Asa goes to Waterbury gets clothes etc. Rainy part of day.

Thursday 18–Rainy all night. Look at cow in morning. Go to Waitsfield am. Work at books & leaches [?] pm. Asa churns & chores

Friday 19–Draw hwd to Waterbury leave it a railroad. Asa works at soap

Saturday 20–Draw hwd to Waterbury. Asa works at soap still

Sunday 21–First meeting at our church this season. Fair attendance

Monday 22–Elwin Templeton came & commenced to work. Mr Seaver works also. Templet on draws hwd to Wby. Seaver hoes corn.

Tuesday 23–Templeton draws horses. Asa & Seaver & oxen work on road. I have calf killed & carry hide & meat to Waterbury. Mrs Wyman & Mrs Smith came.

Wednesday 24–Templeton draws hwd. Rest of us hoe corn. Sell cow to Wilbur etc.

Thursday 25–Asa churns twice. Templeton draws hwd. Finish harrowing lower piece corn to cultivate out upper Price.

Friday 26–Draw potatoes to Middlesex – Boys fix fence & hoe a little.

Saturday 27–Draw potatoes to Middlesex – Boys make fence +c.

Sunday 28–Attend funeral of Mrs. Janus Towle at Moretown.

Monday 29–go to Dowsville am. Templeton draws hwd. Asa & I fix fence & clean cellar

Tuesday 30–I went to Moretown. Get wagon fixed +c am. Grist +c. Templeton draws hwd to Waterbury pm. They draw hwd logs to Geldings am.

July

Wednesday 1–Templeton draws hwd. Commence work on horse barn – Mr. Somerville helps. Wilbur helps at stone work awhile. Warm day.

Thursday 2–Templeton draws hwd. Wilbur hase to hoe, but it is rainy. Somerville sick pm.

Friday 3–I went to Dowsville for plant +c for horse barn. Asa hoes +c. Mrs. Wyman & children went home.

Saturday 4–Templeton draws hwd. Mr. Somerville came & worked on horse barn a little while am. Attend picnic on fair ground am.

Sunday 5–meeting at our ch. Fair attendance

Monday 6–went to Dowsville & got Templeton's colt am. He draws load clapboards to Middlesex & helps to load can finish horse barn. Seaver & Asa hoe.

Tuesday 7–Wm & I go to Ridleys & Waterbury–Templeton draws hwd. Asa & Seaver hoe corn.

Wednesday 8–Go to Dowsville am & sweep [?] hwd boards. Templeton draws hwd to Colbyville. Hoe corn also Asa & Seaver.

Thursday 9–Cut grass on piece above corn with machine. Seaver & Asa hoe. Wilbur [xxxx] part of day, hv mow some

Friday 10–Get in 2 loads hay am. Wilbur & Asa now some below house – rainy pm.

Saturday 11–Templeton draws last of hwd to Colbyville. Asa & I fix hog pen. Asa takes hog to Moretown pm. Templeton & Sissi go to Warren pm.

Sunday 12–Remain at home all day. Templeton & Sissi came am. Rainy in afternoon.

Monday 13–Templeton & I work at saw mill. Draw home saw dust +c at night. Asa churns +c & goes to Moretown for hog.

Tuesday 14–Work at hay. Get in a little below barn as even. Wilbur here part of day. Go to Moretown at eve.

Wednesday 15–cut grass beyond old house & above new barn. Get in 12 loads pm. Fair hay day. Pans came at eve.

Thursday 16–help to get pans am. Men mow north orchard am–on Munson place pm. I go to mill for spouts +c pm.

Friday 17–Good hay day. Now beyond buildings on Munson place.

Saturday 18–nice hay weather. Mow near Munson sugar place. Get in 16 loads.

Sunday 19–Mr Boyce sick. No meeting at our ch. Attend at Moretown.

Monday 20–Go to Waterbury with May & Clinton. Templeton work at mill.

Tuesday 21–Hay weather again. Seaver & Wilbur +c work. Cut last of grass on Munson place & got it nearly all in. Ed Smith came with wife & children. Zihors card at eve night [?]

Wednesday 22–mow above south orchard–boys cart all day. Get in 12 larger loads.

Thursday 23–cut nearly all the grass above road. Very good hay day. Cut above old house & around spring +c.

Friday 24–cut last above –+ swamp piece and Price below ox barn +c. Mr. Wilbur & son go home at night.

Saturday 25 – Ed Smith & family go home am. Cart swamp [?] +c get in 14 loads.

Sunday 26–Sissi & I attend ch at Waitsfield

Monday 27–rainy day. I go to Moretown & to Dr. A Fisks [?]. Asa churns +c. Templeton does not work.

Tuesday 28–Go to Dowsville am. Templeton goes to Middlesex. Asa & I stick up board at Belding mill. Attend funeral of Miss Harriet Turner pm. Mr Tmax came to supper.

Wednesday 29–Templeton & I work at mill. May & Clinton visit at Wm Canerdys & Geo Freeman & to B.J. Stockwells. Asa puts 6 stakes in hay body.

Thursday 30–cut last of lower orchard & piece near graveyard burial ground–get in 5 loads toward eve

Friday 31–now on newly seeded piece below oak & get in 8 loads – not good hay day.

AUGUST

Saturday 1–Templeton draws clapboards to Middlesex & Asa & I work at mill–Seaver cleans barn year +c some

Sunday 2–May, Clinton & I attend ch at Moretown–Rainy part of day. Dr. Hookers child buried today. Died with canker rash.

Monday 3–Seaver & Asa mow piece beyond or barn +c. Templeton I draw hay from Murrays am & work at mill sticking up board pm. Hay day.

Tuesday 4–cut last of grass am. Got it all in but one load pm. Good hay day.

Wednesday 5–Seaver goes to work for Belkey [?]. Asa for Corlisses. Templeton & I draw load from

Murray. I get in one at home am. I work from Wm pm. He goes to Moretown gets shoeing +c for me.

Thursday 6–Templeton draws boards to Middlesex. I go to mill mow away hay +c go to Moretown.

Friday 7–go to Dowsville & mow away hay am & to Moretown & to fair ground pm. Rainy at night. Tom Neill came at eve.

Saturday 8–churn am. Got to Waterbury pm. Templeton draws boards to Middlesex. Asa came at noon fixed fence a little pm.

Sunday 9–attend ch at Waitsfield. May went also. Mr. Dusu [?] preached pm. Felt sick at night.

Monday 10–Templeton draws boards. I churn am. Go to Hoffmans to pick raspberries pm. Warm day.

Tuesday 11–Asa at work for Corliss. Templeton draws clpbds to Middlesex. I go to mill & pick white scattering boards +c am.

Wednesday 12–Judson & I go to Middlesex & load Cxx. Templeton & Wm draw boards to Montpelier. Templeton gets last clapboards from mill. Very warm day.

Thursday 13–Templeton & Wm drawing clapboards to Montpelier. The boards at Barretts. Asa still at Corliss. I do chores & churn twice. Wilbur here. Also Hazeltine.

Friday 14–Asa came home today. He went to Chipman to get oxen shod but did not get it done. I visit at Somerville, am. He & family here pm. Miss J. Somerville, too.

Saturday 15–Asa to work for Corliss. I go to mill & help load ties and churn and carry Sissie to Waitsfield pm. Mary Taylor came back with me.

Sunday 16–Attend ch at Waitsfield pm. Carry Mary back & get Sissi. Warn & pleasant day.

Monday 17–I work at Books. Asa gets cattle shod am & works at fence a little pm. Templeton & Wm draw hemlock to Waterbury

Tuesday 18–Asa & I work at mill sticking up boards & drawing on logs. Templeton draws hemlock.

Wednesday 19–Templeton draws hemlock. I go to Montpelier. May goes also. Visit at ch & ward & do errands at Montpelier. Pleasant time. Asa at Dowsville.

Thursday 20–Templeton draws last of hemlock. Asa drawing bark at Dowsville. I chore & work at books some. [Mrs. Somerville and Belle here pm.]

Friday 21–Asa at Dowsville. Templeton draws wood for Seaver. I churn [chore?] & go to Fayston.

Saturday 22–I go to Greens. Sell him car of lumber. Templeton draws wood from Dowsville. Asa here pm.

Sunday 23–May & I attend ch at Moretown. Sing at Somerville afterward. J. C. Griggs & wife here then.

Monday 24–Templeton & Asa do not work. I go to mill & draw load boards to Middlesex. Buy flour +c. draw logs on logney. [?]

Tuesday 25–I go to mill. Anson goes to Middlesex with boards. Asa is married, does not work. Templeton day not work. I go to Northfield [?].

Wednesday 26–Asa came today. He works at fence & lower. I go to Huntley for Wilbur am. & to mill pm. Ely there to see about wheel.

Thursday 27–I go to Burlington. Templeton at work again. He drives barn [?]. Asa churns & works at Row am. May goes to Wby. I have pleasant time at Burlington.

Friday 28–I stayed at Ed Smith's last night. Asa to work at Row am. Templeton draws boards to Middlesex.

Saturday 29–Wayley of Waterbury came with Eds hors & stayed last night. I went to fair ground to see trot am & to Moretown pm. Make fence. Mrs family here.

Sunday 30–Sissie & I attend ch at Moretown. Templeton away. Asa at home.

Monday 31–Templeton draws boards to Middlesex. Asa works at fence next Elias Marbley. I chore & work at books some.

SEPTEMBER

Tuesday 1–Freeman's meeting. I attend. Templeton goes with me. Harry Bulkly elected Representative. Asa commenced to mow oats.

Wednesday 2–attend John Ingalls funeral. Asa mows oats some. Templeton at horse trot.

Thursday 3–attend horse trot on fair grounds at

Waitsfield. Asa mows oats. Templeton does not work.

Friday 4–go to Corliss am & to Middlesex to load car afterwards but got no car. Asa mows oats.

Saturday 5–Boys finish mowing lower piece oats. Draw in 2 loads at night. I churn & help about oats.

Sunday 6–camp meeting at Northfield. Do not attend ch anywhere.

Monday 7–Mr. Seaver mows oats in south orchard. Asa does not work. Templeton at work with oxen on road at Wilbur.

Tuesday 8–Asa at work again. Templeton at Wilbur. Asa & I make fence am. He & Seaver now pm. I go to Moretown for Earl & family at night. Christine & family at Wms.

Wednesday 9–Earl & XXXXX & Wm & families here on visit. Men mow last of oats & we get in 7 loads pm. Templeton at Wilbur.

Thursday 10–churn am. Attend fair pm. Earls People May & Sissi & Berlin then also–Templeton broke wagon at night.

Friday 11–I draw boards to Middlesex. Have Somerville's wagon. Templeton does not work. Get in last of oats pm.

Saturday 12–Templeton & Seaver dig potatoes. Asa draws manure. I visit with Earl–go to Moretown pm.

Sunday 13–remain at home all day.

Monday 14–Thrashing machine here. Earl & family go to Wms at night. Seaver digs potatoes.

Tuesday 15–Finish thrashing & move machine to Wms at night. I go to Crandall eve. Earl came at night.

Wednesday 16–I go to Dowsville & see Wilbur road. Measure bark +c am. Mary Taylor here pm. Templeton draws boards. Asa at Wm Seaver churns +c.

Thursday 17–Asa at Wms. Earl, Tem[pleton] & I go to Middlesex & load car. Seaver chores a little.

Friday 18–Seaver chores a little. Asa works on old house. Templeton away pm. I settle with Wilbur am. Went at Mr. Somerville with Earl & family pm.

Saturday 19–I go to Moretown & to Munson–go to Waterbury afterwards. Templeton works with oxen. Asa does not work. Seaver chores a little. Earl & family go to Wm.

Sunday 20–May and I go Waitsfield with Earl's people & to ch. Rainy pm.

Monday 21–I fix fence & am go to Wilburs to see XXXins land. Wm goes too. Asa does not work. Seaver does not work. Templeton digs up potatoes.

Tuesday 22–Templeton draws load hwd to Waterbury. Earl & family go to Ch[ristine?]. I take them to Wby & go to settle with Elliott & Co. Asa digs potatoes.

Wednesday 23–I draw hwd to Wtrby. Boys dig potatoes. I went to Moretown for millwright, but he was not there.

Thursday 24–Boys dig potatoes. I churn and can help about potatoes pm. Go to Moretown for Rumwell at am. Cow died pm.

Friday 25–I work at mill. Templeton draws hwd to Wtby. Asa cuts corn.

Saturday 26–I work at mill. Templeton draws hwd. Asa and Seaver cut corn some on lower piece.

Sunday 27–Rumwell, Sissi & I attend ch in Waitsfield.

Monday 28–work at mill. Templeton does not. Asa & Seaver cut upper piece corn.

Tuesday 29–Asa & Seaver cut corn on lower piece. I churn twice boil swill for hogs & bind corn & got in one load at night. Templeton draws hwd to Wtby.

Wednesday 30–Templeton goes to Moretown gets wagon fixed. Asa & I husk corn. Seaver churns.

October

Thursday 1–Templeton draws lumber to Waterbury for organ for ch. Men work on potatoes. Went to Waitsfield am–to Moretown eve.

Friday 2–Templeton draws load organ lumber for Crandall. Seaver, Asa & I dig potatoes. Sell corn to mare [?] with Seaver.

Saturday 3–Men dig potatoes. Templeton draws hwd to Waterbury. I go to Fayston & to Gilbert pm. Ray Pigs [????]

Sunday 4–attend ch at Moretown

Monday 5–Seaver digs at his potatoes. Templeton draws lumber to Waterbury & gets flour. I go on mountain to let job cutting logs.

Tuesday 6–Templeton through. I pay him off. Asa picks 2 bush[els] apples am. Get in corn pm.

Wednesday 7–Election Day–Asa at Montpelier–I husk. Seaver digs his potatoes.

Thursday 8–Pick apples on Munson place & in S orchard

Friday 9–Pick apples some. Asa visits with Davis–go to Mr. Geo Hills for Charlotte eve. Rainy some.

Saturday 10–Carry Sissi to Warren. Asa picks apples in N orchard.

Sunday 11–Charlotte & I attend ch Moretown.

Monday 12–Langdon Davis come to work. He helps Seaver. I husk corn.

Tuesday 13–Davis helps Seaver. I go to Moretown & to mill. Get in corn in morning.

Wednesday 14–Husk 44 bushels corn. Asa & Davis husk. Seaver draws manure. Attend wedding at Mr. H. Griffith in eve.

Thursday 15–go to Dowsville & Moretown +c. Engage Canerdy [?] & wife to go on mountain. Men draw corn and pick apples.

Friday 16–Husk corn. Asa does also. Seaver banks house. Asa and I make road on hill am.

Saturday 17–Husk corn. Seaver goes to Moretown & gets oxen shod & goes to mill.

Sunday 18–Attend funeral of Mrs. H T Campbell Fayston

Monday 19–kill hog am. Mow on hill pm Go thru Dowsville & get Canerdy's people.

Tuesday 20–Skid logs on mountain. Get along well. William Montgomery commenced to work at noon.

Wednesday 21–skid logs. Buy logs of Parchu & Stockwell in eve. Asa draws wood.

Thursday 22–skid logs. Asa drawing wood.

Friday 23–came from hill. Ox lame. Husk corn.

Saturday 24–William husks corn. I go to mill and work. Asa draws wood.

Sunday 25–Attend funeral of Stockwell girl.

Monday 26–Attend funeral of Folsom child. Finish husking corn. Take apples to cider mill in eve. Tom Neill here Argan came

Tuesday 27–I go to mountain but do not stay. Draw wood to house some. Asa thro work. Tom Neill gone home.

Wednesday 28–Draw wood am. Go to mill with Belknap pm. Mrs. Kellogg taken quite sick.

Thursday 29–William plows a little on piece below sugar place. I carry May to Waterbury.

Friday 30–William plows some. I go to Moretown.

Saturday 31–Asa Corliss moves away. William plows some I chore am & go to Wtrby for May pm.

NOVEMBER

Sunday 1–Charlotte & I attend ch Moretown.

Monday 2–William chores & plows a little. I go on mountain & skid logs. Stop at Canerdys at night. Canerdy works.

Tuesday 3–Skid logs on mountain. Wilbur cuts roads. Rumsill came to fix mill today.

Wednesday 4–finish skidding logs & move home. S Somerville's mother dead.

Thursday 5–Wilbur finishes plowing piece below sugar place. I go to Wms am to work at books and write letters pm.

Friday 6 –Wilbur Annington and Pero here to settle up for cutting logs. Spent half day with them. Got cider from mill ec. Wilbur draws stones a little.

Saturday 7–Rumsill here–go to mill with him & fix it. Wilbur churns and plows a little – nice warm weather.

Sunday 8–Charlotte and I attend ch in Moretown. Mrs Somerville here to sing at eve.

Monday 9–Finish plowing piece on Munson place and commence small orchard piece. Wilbur draws stones

Tuesday 10–Take cow to Greens pm – plow some, Wilbur draws few stones, Seaver, Asa, Wilbur and Templeton here

Wednesday 11–Go to Wms and to mill to Chipmans and Moretown. Get bands repaired.

Willie draws load or two of manure. Get in garden sower. [?]

Thursday 12–Go to mill with bands Put aquaduct +c. Draw home load dust +c Willie chores Cold

Friday 13–I churn twice am Willie chores Get in sugar wood pm Cold

Saturday 14–Go to mill with grain Willie chores

Sunday 15–remain at home all day

Monday 16–go to Dowsville and put logs on logway Go for Lizzie at eve Willie draws few loads manure

Tuesday 17–I go to Montpelier May goes too Willie does chores and draws couple loads manure

Wednesday 18–plow in small orchard – Willie draws manure. Wilbur–Donnovan–Seaver–Templeton–Jas & Charles Whelan here

Thursday 19–Willie does not work I do chores & shoe bobsled

Friday 20–Willie churns am. Draw sugar wood and hay to horse barn Put hay bodies under cover +c.

Saturday 21–chore am Carry Charlotte Hills home & go to Waterbury pm. Willie chores.

Sunday 22–Remain at home

Monday 23–Stormy. Commence to oil harnesses am. Kill hog & go to Wtrby pm. Fix water pm

Tuesday 24–finish oiling harnesses Willie draws wood some

Wednesday 25–I work at sawmill Willie does chores & draws wood some

Thursday 26–Thanksgiving Day. Draw wood am, visit at Somervilles pm. Wyman and family out

Friday 27–Willie draws wood a little. I go to Moretown am get colt shod. Mr and Mrs Gilbert & girl Mr Canerdy & family Mr & Mrs Somerville here

Saturday 28–Draw wood work at books +c. May & I go to Wms at eve

Sunday 29–May sick Dr Hooker here at eve

Monday 30–put harness together am weigh hay for Seaver +c. Go to Moretown & mill & get horses shod +c. Willie draws wood.

December

Tuesday 1–go to Dowsville and get out hemlock logs from Wilbur. Draw logs on logway & stick up boards half day Wilbur and Templeton work

Wednesday 2–work at Wilburs logs come home eve – Cow sick

Thursday 3–finish getting out hemlock & commence on spruce of Miners

Friday 4– draw spruce from Miners Wilbur & Templeton still work

Saturday 5–finish Miners logs & move [?] home at night

Sunday 6–remain at home

Monday 7–went to Waterbury Willie went to Chipmans & got oxen shod

Tuesday 8–went to mill & drew on logs & stuck up boards Wilbur worked Draw logs to Ed Belding

Wednesday 9–attend funeral of Mrs Strong of Fayston–go to Neills & Blomers afterwards

Thursday 10–go to Wms & try to settle with him. Do his chores +c. Suce & Geo Thornton there – visit at Somerville at eve Holberg family there

Friday 11–go to Wms in morning Work on books at home afterward – sell butter and sugar to Seaver and buy potatoes of him

Saturday 12–R from school. He [?] family here friend. Visit with them. Hog sick. ober & also Somerville people pm. Sing school eve [Ed note: none of this entry is clearly legible]

Sunday 13–remain at home. Willie at Dowsville, hog died this am

Monday 14–draw load potatoes to Middlesex. Cold

Tuesday 15–Very cold. Go to Wms pm Willie chored. Go to Moretown am

Wednesday 16–draw potatoes to Middlesex Willie chored

Thursday 17–Chore a little

Friday 18–Go to Waterbury buy goods of Weyman. [Next line unintelligible]

Saturday 19–choir am. Put glass in window +c. William here pm. Go to Dr Kingsley to sing in eve

Sunday 20–May Clinton & I attend ch in Moretown

Monday 21–I go to Moretown am, work at Dowsville pm, draw logs +c

Tuesday 22–draw logs at Dowsville get along quite well. Draw board & slick up to draw logs on logway. Willie works also.

Wednesday 23–work at Dowsville drawing logs. Willie sticks up boards +c – come home at eve

Thursday 24–go to Wms am & Waterbury pm. Attend Christmas celebration at Moretown in eve

Friday 25–chore +c am. Carry Lizzie to Waitsfield pm. Attend school meeting & sing school eve

Saturday 26–went to Moretown am & to Dowsville pm. Ezra Smith & Eva Wyman here in pm & in eve

Sunday 27–May Clinton & I attend ch in Moretown

Monday 28–go to Moretown with butter & get cattle am. Churn twice pm. Willie went to Dowsville with horses. Rainy pm

Tuesday 29–go to Wms am, split wood +c chore pm squally pm Mr Gillett here pm

Wednesday 30–cold & windy, split wood some Remain at home. Mr Gillett stayed the night – go to Somerville eve

Thursday 31–go to Greens & Moretown in am. Wilbur here pm. Weigh hay & go to Dowsville. Cold weather

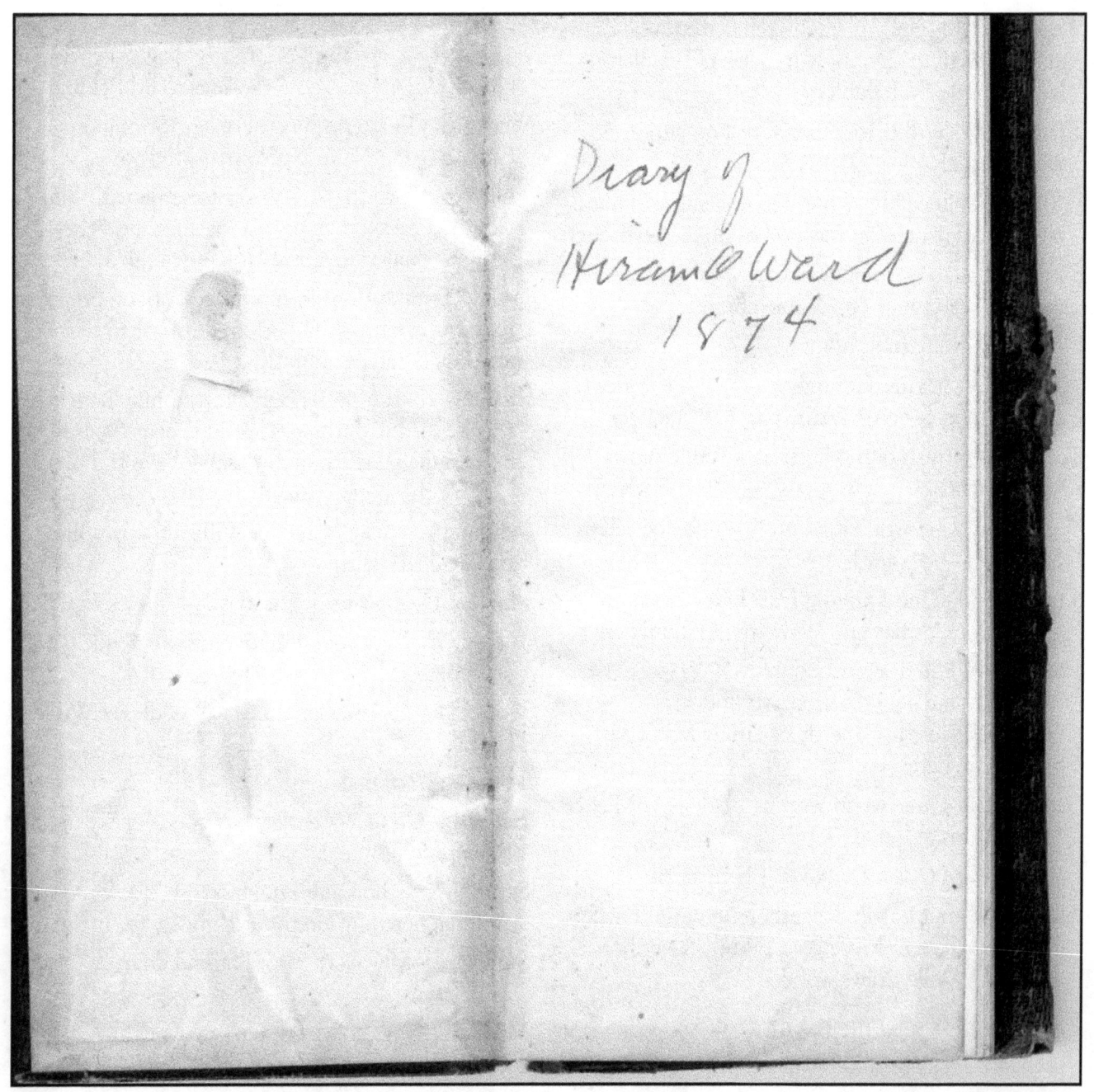